MW00974239

RADIO FREQUENCY IDENTIFICATION APPLICATION 2000

By

James D. Gerdeman

RESEARCH TRIANGLE CONSULTANTS, INC.
Cary, North Carolina

Radio Frequency Identification is the second in a series on enabling technologies published by Research Triangle Consultants Inc.

©

Research Triangle Consultants, Inc.
1135 Kildare Farm Road, Suite 329
Cary, North Carolina, 27511
Phone: 919 - 469 - 5111
Toll-Free: 1 - 800 - 382 - 5284 (USA)
FAX: 919 - 380 - 1727

Editor: Joanne LeRose
Assistant Editor: Anthony Concia
Layout & Graphics: Joanne LeRose
Printed and published in the United States.

ISBN 1 - 883872 - 01 - 4
1. Radio Frequency Identification. 2. Title

DEDICATION

To my family, especially Mary Jo, Jim, Susan and Megan

ACKNOWLEDGMENTS

There are a number of individuals, companies and organizations that have helped to inspire and support this project and have been so helpful in many ways. As always, my customers are my number one supporters. Primary support has come from my friends who worked in the Automatic Equipment Identification department. They have given me the strength to want to produce this book. Joe Burnam, Stan Strickland, and Mark Choberka deserve just credit. My supporting cast, Marty Lamia, Marty Clarke, my business partners and friends, my business family, and all of the Associates have been very supportive. My own family has helped as well; my brothers, Dave, Bill, Alan, and Mark, family support from Don and Ceil, Norma and Joe, my son, Jim, and my daughters, Susan and Megan. But, if Mary Jo had not supported this effort, the ink would never have made it to paper. Thanks to Mary Jo and all of you!

ABOUT THE AUTHOR

Jim Gerdeman has used his Industrial Engineering Degree from the University of Dayton to help him focus on customer applications. As a computer systems engineer, he has developed systems to improve customer operations in a number of industries including banking, communications, manufacturing and transportation. He managed the complex solution development for the transportation industry team of one of the largest computer manufacturers in the world. He has written technical papers that describe the use of Radio Frequency Identification and has published articles in various trade journals. His attentions have been focused on the transportation market with the use of RF/ID for Intelligent Vehicle Highway Systems, Electronic Traffic and Toll Collection Systems, and Intermodal Container and Railroad applications. He is well versed in the advancing technologies for transportation, and has spoken world-wide at conferences on leading edge technologies.

PREFACE

Identification of things is one of the most important functions we do in our industrial and public societies. Radio Frequency Identification provides us with the capability to identify things with the least amount of disruption to the main activity being performed. Vehicles can be identified as they are driven by toll gates. Rail cars can be identified as they roll by a track side. Trucks and buses are easily identified. People can walk through a gate without reaching into a pocket for an ID card. Packages can be identified without the need to orient a side towards a reader or scanner. In addition to the ID, Radio Frequency Identification Systems offer the capability to provide additional operational data. All of this and the RF technology affords the user increased accuracy and real-time data.

If you intend to read only the sections of this book that obviously apply to you, then you will miss a lot. The idea is to cover as many of the applications as possible after setting the stage with some of the fundamentals. The problem with skipping a section relates to how you have your companies and organizations function. You are multi-faceted. You might be a railroad, but you have gates for employees, vehicles, draymen and others. You might own a fleet but also have a rail siding. It is recommended that you read through the entire book before concentrating on particular sections.

There are a number of pictures and diagrams to assist you in the understanding of the concepts. I believe a picture is worth more than a thousand words as the saying goes. It was not my intention to specifically discuss each picture or diagram. They are provided for your review and study.

The first half of the book is dedicated to general radio frequency topics. Specific applications are presented in the chapters; Toll, IVHS, Air, Rail, Intermodal and Fleet. The final chapters cover Justification and the leading RF/ID manufactures.

TABLE OF CONTENTS

CHAPTER I

OVERVIEW

Customers, managers, and operations personnel have long desired to improve the accuracy and timeliness of data collection. The costs associated with the equipment used in our day to day operations and the sheer number of items to keep track of can boggle the minds of the best staffs that money can buy. Every day millions and millions of things move from here to there. Vehicles leave their gated communities and condominiums. Cars travel from toll gate to toll gate. Roadways take workers to and from their places of work. Trucks enter and leave freight terminals. Delivery vehicles deliver. Packages move from warehouse to store; store to customer; customer to disposal facilities. While all of this is going on there are computer systems to control the movement, processes and identify the objects or equipment each step of the way. Bills of lading are printed and annotated and copied. Traffic counters count and reports are generated. Delivery persons deliver and consignees sign on the dotted line. Computers even capture the signature electronically.

To improve the accuracy and timeliness of identification we have used the bar code. Bar code wands read labels on containers, pallets, boxes, and items. Identification is so important to our business processes that we build our processes around the ability to read the bar code labels. Bar codes have helped us increase accuracy and decrease the cost of handling the items. In fact some processes use bar codes to automate the handling and routing functions. Why is all of this so important?

There is the ever present changing demands of the customer and the increasing cost to process the order. Labor costs continue to rise. The cost of equipment and facilities are high. There are never ending pressures to be more productive and to get more from the assets of the company. Where is the asset and who is using it? The customer is demanding more information in their service requests. They want just-in-time information and delivery. They want to be told when the plan is out of sync and delivery commitments cannot be made. The customer is willing to pay for this service. If you can improve the time to get through a gate, the customer will invest in additional equipment to get through a toll road. As the world gets smaller, anything to assist in the international flow of goods becomes important. It becomes more and more important to identify equipment and the people who handle the equipment.

Bar codes have given us significant improvements to identify the people and things in the process. But as the applications expand, there are two fundamental road blocks that prevent progress from taking giant leaps. These road blocks are fundamental to the solutions that would yield automatic and accurate data collection.

- The processes tend to be manual.

- The environment and distance requirements are limiting.

PRODUCTIVITY

TIMELINESS

ACCURACY

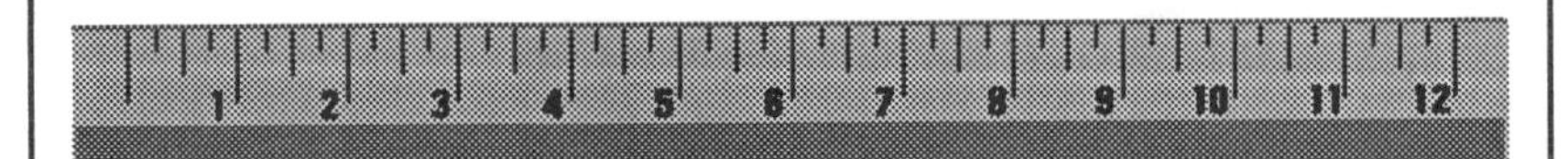

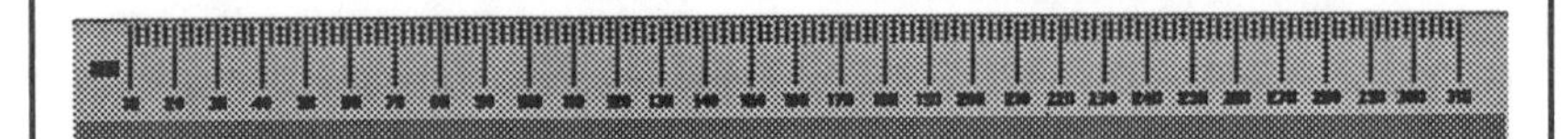

These fundamental concepts have caused us to search out another form of identification technique. Before the bar code enthusiast gets too defensive, it is a fact that bar code identification will be here for a long time and has been one of the more significant technologies. In fact, the economy of the less than a penny bar code label will out perform the often multi dollar cost for radio frequency transponders for a long time. However, when the application requires increased accuracy and improved timeliness of data, and, requires significant productivity in the manual process and an outside environment with longer read distances, the radio frequency transponder will win out over bar code.

MANUAL PROCESS

"Eighty thousand times a day, a long shoreman takes a dull pencil and writes on a soggy piece of paper the ID of a container to be key entered later..." This statement is from a friend of mine who is very active in the shipping industry. This process is fraught with opportunity for error. The identification can be misread; letters and numbers can be transposed; and again, the key entry operation can be misread and, numbers and letters transposed. If the identification is also associated with a location, the process might not be fast enough to give the computer system the information. These records are generally placed into batches. The batch process takes time. Key entry clerks take coffee breaks and have other interruptions, so time elapses. The equipment might have been moved before the original location can get into the system. 85% accuracy is good.

In another example, let us use the bar code system to increase the accuracy of loading parcels onto a truck. The parcels have the bar code and the material handling personnel have a hand held bar code wand. This is a good example of increased accuracy of identification and there are many advantages possible, such as verification of destination and truck route. Reduced misdirection of the packages is possible.

The process of reading the bar code on the parcel involves several steps. The bar code must be in view, so the package must be oriented with the bar code easily accessible. The bar code gun then needs to be pointed at the bar code so that the identification can be made. If the package is large enough then the material handler will have to place the package on something like a table, or floor, to free a hand and then make the bar code ID. This all takes time. Some engineering has been accomplished to make special bar code readers to identify the packages but there is always an orientation problem. If the label is out of view, there is no identification. This takes time.

Here is one more example on this for now. I visited an airport where the staff was so proud of a new system they had installed for controlling the limousines and shuttles. It had a very reliable and accurate reading mechanism of a magnetic stripe. This was a card similar to a credit card. The driver would stop and roll down the window, reach out and insert the card, and a gate

PHOTOS BY GERDEMAN

would open. This was a very nice system and had significant controls in place. The back office reports on frequency of loops through the airport and other information was very impressive. The problem, of course, was that the vehicle had to stop, and there was the inconvenience of the rain and cold on the driver's arm. These processes, while providing improved accuracy, have additional manual steps that could be eliminated to provide for increased productivity.

ENVIRONMENT AND DISTANCE

As more and more applications are thought of, in the quest to remain competitive, or to solve the customers strategic wants and needs, there is a demand to operate in/out doors and at significant distances. These both have been beyond the specifications for bar code.

The railroads and the intermodal community tried using bar codes to identify rail cars and containers. In fact, there was a fairly high read rate for the special bar code labels. But, overcoming the problems of reading a bar code in the elements proved to be too costly for the much needed ID system. Rain, snow, and dust storms all prevented or severely impacted the read rates for the bar codes. Fog is a problem as well. The elements presented a basic reliability problem for the bar code reading system.

It turns out that there continues to be steady improvements in the ability to read bar codes. Media assisted bar codes improved the reflective characteristics. More processing power enabled routines to increase the read range. Several feet is generally accepted today. This is a significant improvement over the inches required in the past. Nevertheless, users continued to think in terms of a football field away. Applications for 20, 30, 75 feet began to surface.

As the Radio Frequency Identification (RF/ID) system began to evolve, the user community could begin to throw out the old paradigms and begin to think of new opportunities. With a radio frequency transponder and the associated reader equipment, it became possible to read ID's while driving past the containers in an intermodal yard. These yards can cover several square miles and hold thousands of containers. Instead of thinking of counting the items on a specific shelf in a specific aisle of a specific store, the idea of counting all the items in the store became possible. These examples show a new magnitude of possibilities. To place a framework around our possibilities, let's focus for a few minutes on the things that are to be identified, and the application areas of concern.

There are millions and even billions of things to identify. There are a number of ways to look at the classification of things to identify, just to get a perspective. There are big and expensive things, there are small and expensive things. (*see TABLE: THINGS TO IDENTIFY*)

TABLE: THINGS TO IDENTIFY

ID TYPE	DESCRIPTION	APPLICATION	CUSTOMER	POTENTIAL
BIG & BULKY ASSETS	CONTAINER ASSETS cars chassis containers locomotive rail cars refers trailers ULD's	GATE YARD SERVICE CONTROL	Where is? How used?	$600 million
		IVHS TOLL	How used? collection	$2 billion
PALLET & LARGE	SHIP IN CONTAINER pallets over 50# large tires engines boxes	DOCK GATE	Where is?	$500 million
Package	ENVELOPE letters overnight reports small box	SORT	TRACKING	$1 billion
		ID	SALES	$10 billion
PEOPLE	employees drivers students vendors	DOORS PARKING GATES	SECURITY	$1 billion

NOTE: This table is not all inclusive. The dollar amounts are gross estimates of annual sales.

TABLE: Things To Identify shows us a type with a description. The table highlights some of the applications, what the customer might be interested in knowing and the potential dollar estimate.

BIG AND BULKY ASSETS

Big and bulky assets are those that are valuable as well as expensive. A trailer or container might cost several thousands of dollars and a locomotive may cost a half a million dollars or more. These assets are expensive, and need to be utilized and cared for in a way to maximize their usefulness. The following table shows some examples of prices from a used equipment catalog.

SAMPLE USED EQUIPMENT PRICES

EQUIPMENT TYPE	LOW PRICE	HIGH PRICE
Automobile	3,500	49,000
Bus	25,000	69,000
Crane	18,900	58,900
Dolly	850	3,250
Dump Truck	7,900	18,950
End Dump Trailer	18,000	28,000
Flat Bed Trailer	2,450	13,750
Fork Lift	9,500	14,000
Gen Set	10,000	15,000
Locomotive	350,000	1,200,000
Pickup	7,450	19,500
Rail Cars	7,500	120,000
Reefer	4,950	13,000
Reefer 48 Ft.	12,950	25,850
Stake Body 14 Ft.	8,500	15,000
Tank Trailer	6,700	22,500
Tractor w/o Sleeper	16,500	39,500
Tractor w/ Sleeper	16,500	58,500
Trailer 48 Ft.	7,000	10,500
Van 15 Ft.	12,000	13,950
Van 20 ft.	14,850	20,000

These prices have been generated from used equipment catalogs and classified ads. The example amounts are in dollars.

TABLE: SAMPLE USED EQUIPMENT PRICES

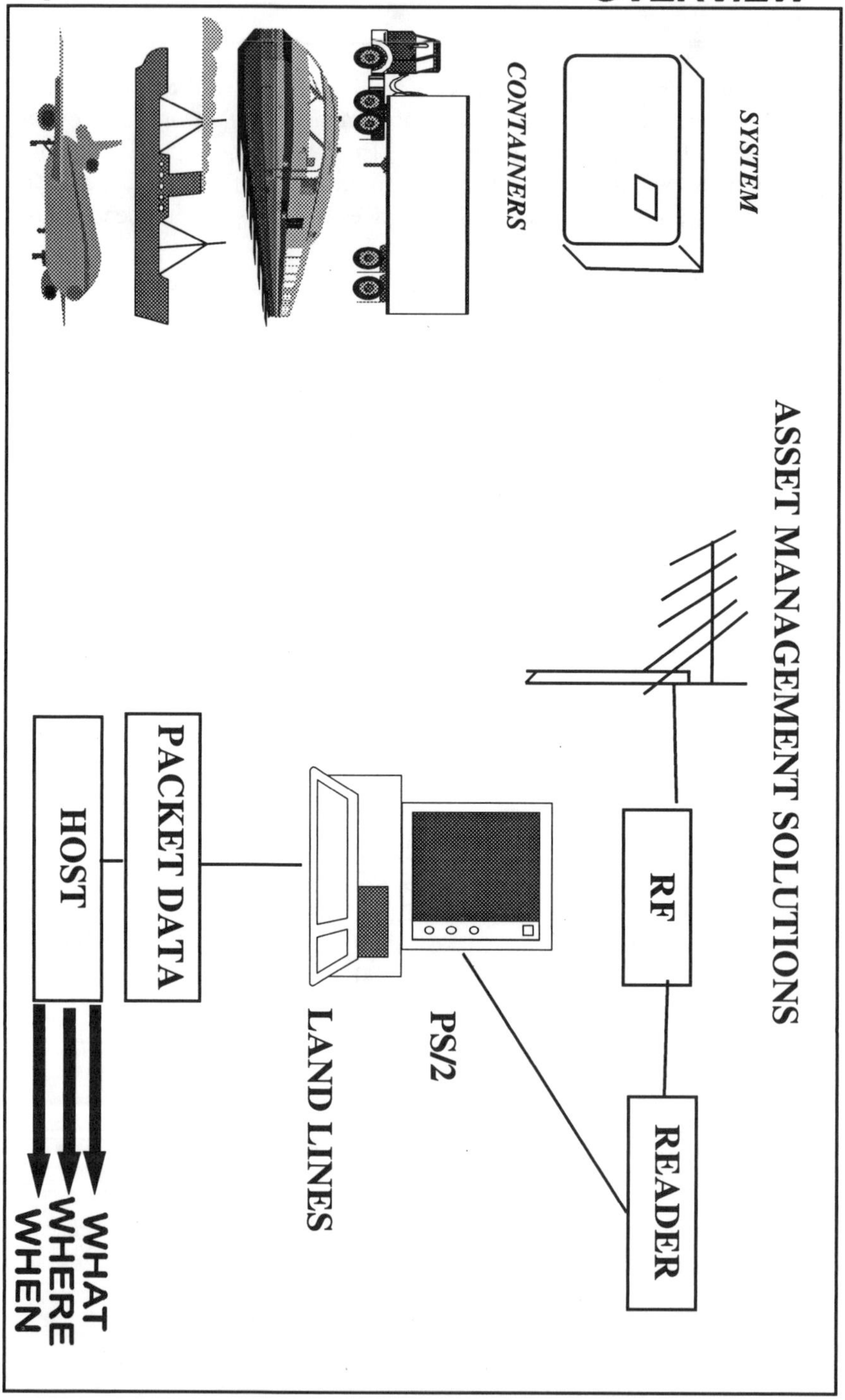

ASSET MANAGEMENT SOLUTIONS

SYSTEM

CONTAINERS

RF

READER

PS/2

LAND LINES

PACKET DATA

HOST

WHAT
WHERE
WHEN

With costs like these, it is easy to have several million dollars of equipment invested in a business. Equipment cost and the maintenance involved is second only to personnel costs in most profit and loss statements. Understanding how the asset is being used, along with where the asset is at the current time, peaks the interest of operational management.

Information made available to the management team, including a reliable and timely identification, provides an interesting set of strategic solution options for a business. A study recently completed showed that the freight transportation business has over 61 million assets to be tracked and maintained. It was estimated that 16 million assets had been owned by companies who could afford the support costs for the estimated systems necessary to take advantage of the new technology. The estimate for the identification system potential in this classification is $600 million.

The estimate used to calculate toll road and Intelligent Vehicle Highway Systems, (IVHS) is even more significant. The two billion dollars for the radio frequency identification includes applications of toll collection and traffic flow. These applications will tend to bring the public, as well as the transportation groups, into the equation. The affluent will pay a little more to use an Automatic Vehicle Identification (AVI) system, to pay tolls in the high speed lane. Car pools will be given radio frequency tags to identify and validate the use of the high speed lanes. The radio frequency tags will even turn traffic lights from red to green for the car pool vehicle at special use lanes of on-ramps. Traffic departments will use the radio identification for traffic flow indicators.

So, for the big and bulky assets there are two themes. The first is directed at the use of the asset. Where is it, and are we using the asset properly? The second is directed at using the big and bulky asset on the infrastructure of the highways and byways. Both are significant and strategic problem areas.

PALLET AND LARGE

These are items that must be counted that fit into containers described in the big and bulky section. These items are pallets, items over fifty pounds, bulky tires, and large boxes. These things are transported in many ways and by many carrier types. The idea is to track the asset through the process. This means identification as it flows through each loading dock. This area is thought of as applicable to less than truck load carriers and special service carriers. It is estimated that the use of radio frequency identification in this process has a $500 million potential.

PACKAGE

Package delivery offers significant potential for the identification market place. This potential certainly involves the objects that are delivered by the air freight industry and the U. S. Post Office. Certainly, the cost factors are providing additional technology pressures. The facsimile business and electronic mail offered in computer systems are causing additional package reduction for this classification of items. Still, one over-night package delivery company delivers over one million packages per day. In addition, during the peak of the holiday season another company delivers 14 to 18 million packages per day. This kind of volume, and the need to deliver a physical item in a short period of time, provides an estimated $1 billion of opportunity for tracking applications.

The hidden potential is the package, as most of us know from our visits to department and convenience stores. The shelves are full of the items we love to buy. These items include mixers, radios, TV's, diskettes, tapes, saws, drills and more. You get the idea. Store management often budgets significant dollars to perform a physical inventory. As the technology advances, it will be possible to do this within seconds with the radio frequency applications of the future. The potential is huge, $10 billion. The idea has surfaced, and the technology is available, but needs some refinements.

PEOPLE

Perhaps one of the best opportunities is to add the radio frequency identification to the personal identification schemes we know today. Employee badges have had the bar code, the magnetic stripe, and the Smart Card, so why not an RF transponder? This is not a new idea. RF/ID systems have been in use for several years and have been a reliable identification method. The employees do not even have to remove the ID tag from their wallets and purses. The tag is read as the employee walks through the doorways fitted with a reader system. When used in combination with other security techniques, the potential of $1 billion will be realized. The same type of ID can be used in parking lot and other gate applications. Shared parking lot schemes will assist in the utilization of scarce parking resources.

SUMMARY

The application potential for radio frequency identification is based on the need to eliminate manual processes and the need to read the ID in harsh environments at long distances. This will assist operational management streamline the processes of handling the millions of assets. There is significant potential to be realized in the RF/ID market place.

CHAPTER II
FUNDAMENTALS OF RF IDENTIFICATION

There are several key components to the radio frequency identification system. The transponder or tag is placed on the asset or thing to be identified. The antenna and reader components provide the process for reading the identification on the tag. Special equipment is used to place the identification on the tag. These components provide the desired functions for the system.

The primary objective of this section is to review the components of the radio frequency identification system and the user functions that may be offered by various vendors of RF/ID equipment. Within this context a functional analysis is required so that we can place the components into the proper context.

As was previously discussed, the identification system is to provide an accurate and timely identification. This fundamental objective of the RF system is to provide the justification for the implementation of the system. Functions to be performed range from the actual identification to the distribution of the identification data.

BASIC RF/ID EQUIPMENT

There are four fundamental parts to the radio frequency identification system. These are: 1. Transponder 2. Antenna 3. Reader 4. Encoder

TRANSPONDER

The transponder is a relatively small device that contains the identification code. A transponder is often referred to as a tag or emitter, and sometimes as a label. This device is available in many different shapes and has a variety of memory options. Transponders are available in read-only, dynamic, and read / write options. Transponders can be passive. This means that it is powered when it encounters a interrogating signal. An active transponder is one that has a battery and that is continuously transmitting.

ANTENNA

There is an antenna at each end of the RF/ID system. An antenna is a metallic apparatus for sending and receiving radio waves. The tag has an antenna to send and receive data to and from the reader system. Antennas come in different sizes and shapes, and they also produce a specific signal pattern in which to engage the communication. Some antennas are high range or long distance while others produce short patterns. They have various mounting options.

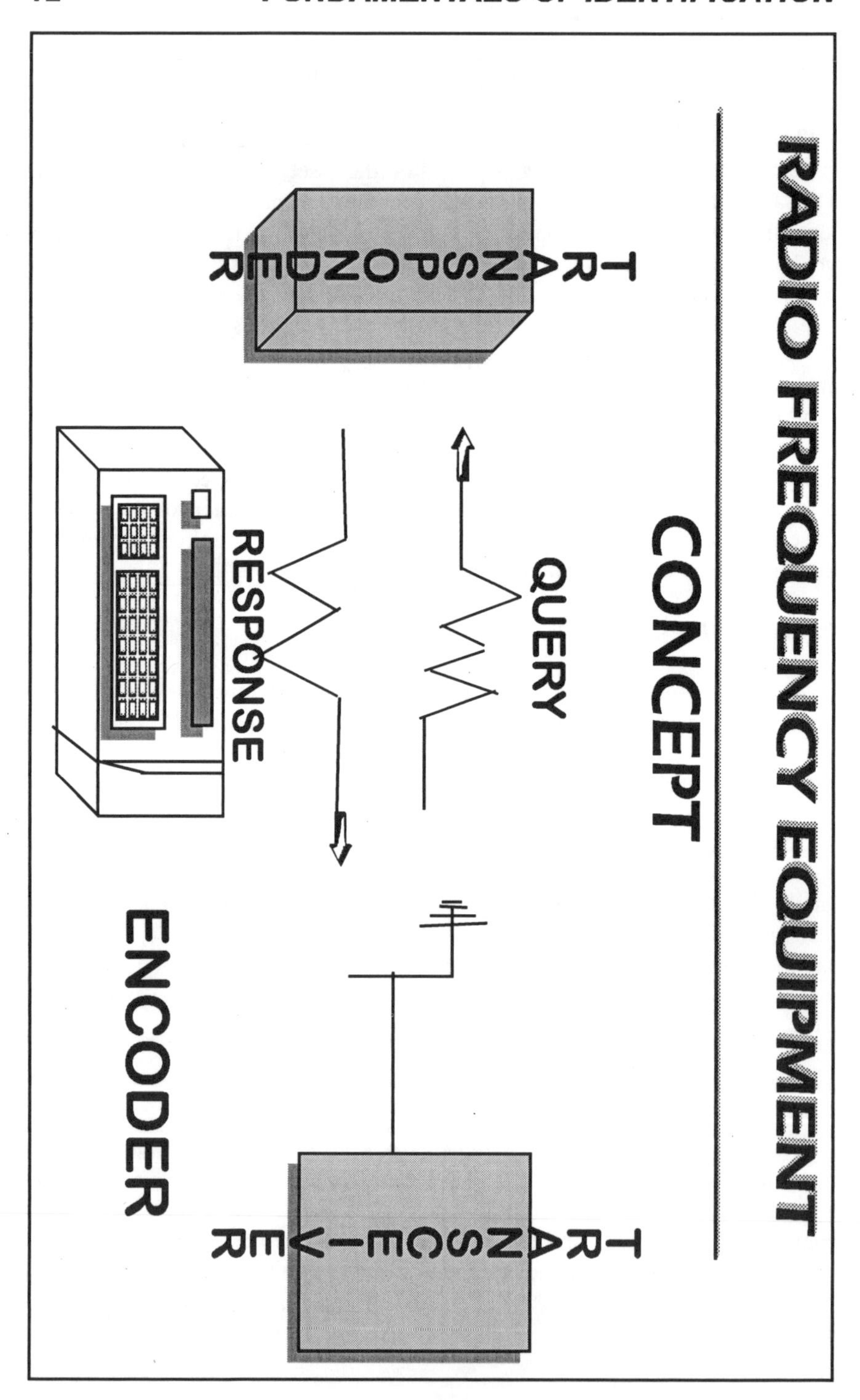
RADIO FREQUENCY EQUIPMENT
CONCEPT
TRANSPONDER
QUERY
RESPONSE
ENCODER
TRANSCEIVER

ANTENNAS

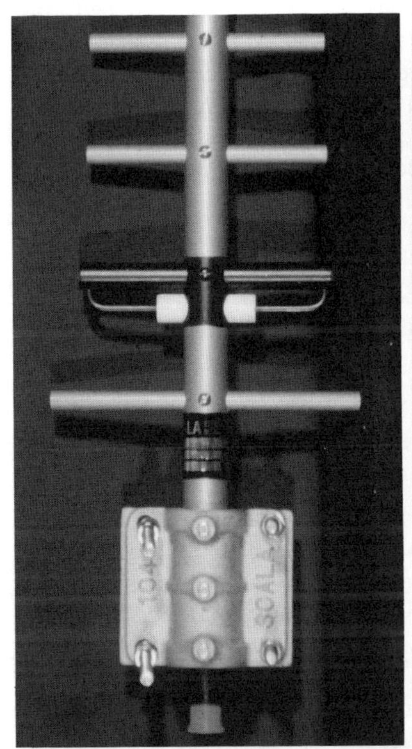

READER

The reader device interprets and decodes the radio signal and performs other system functions. Typically, a time stamp is added to the decoded radio signal and information from input and output devices are added to the reader's memory. Reader devices tend to range in function from simple interpret and decode to more sophisticated control devices. These devices also provide the command streams to write to the transponders in the read / write systems.

ENCODER

An encoder is a device used to place the identification code on the transponder. This function is generally thought to be permanent and a one time operation. The identification data is encoded or written to the transponder.

HOW THE RF/ID SYSTEM WORKS

Simply, the process would work like this. A user would take a tag and place it on the encoding device. The user would then key enter the codes for the desired identification. The user would then place the tag on the object to be identified. Even though the tag could be mounted in a temporary fashion, it is usually mounted to the object permanently, so that it can be read identifying the object.

The user would place the reader at strategic points. When the objects with the tag pass by the reader, the tag is read and by association the object is identified. The system uses a special radio frequency technique that is especially advantageous for motion and distance applications. As the tag passes by the antenna, it reflects back an identification signal. The reader receives the reflected signal and processes it to determine the identification code. The reader then relays the ID to a computer.

NON-CONTACT ID

The wonder of RF/ID is that the media of the identification code does not need to be in contact with the reader mechanism. In other words, there can be a separation of the identification code media and the reader control system. In fact, there can be several inches to many feet separating the two. One consultant uses the term, "a football field away". This advantage makes the identification a hands free system from the reader control point of view. There is a communication that is needed and depending on the type of system and the manufacturer of the RF/ID system, the communication technology employed might be different. They use radio frequency. What is radio frequency? There is a spectrum of energy description that might help put it all together.

Frequency ranges include power, radio, radar, light, X-ray and so forth. Different parts of the spectrum have been used for identification systems in the past. The following chart shows the frequency ranges.

FREQUENCY RANGES

	10^{22}	
	10^{18}	**X-RAY**
	10^{16}	**ULTRAVIOLET**
	10^{15}	**VISIBLE**
INFRARED ID SYSTEMS	**10^{14}**	**INFRARED**
	10^{13}	
		MICROWAVES - RADAR
	10^{10}	
HIGH FREQUENCY RF/ID		
	10^{9}	
	10^{8}	
		TV - FM RADIO
	10^{7}	
		AM RADIO
	10^{6}	
	10^{5}	
LOW FREQUENCY RF/ID	**10^{4}**	**INDUCTION HEATING**
	100	**POWER - DC POWER**
		AUDIO - (HUMAN EAR)
	0	

CHART: FREQUENCY RANGES CONCEPT

FCC CODES

According to the Federal Communications Commission, there is a range of frequencies that are available for use. The allocation of broadcast frequencies is controlled by the FCC, and the use of the various frequency ranges have been designated by the FCC. The bands of 902 to 912 and 918 to 923 MHz have been made available for Automatic Vehicle Monitoring, (AVM). The AVM systems can operate in these frequency ranges providing they do not interfere with Government stations operating in these bands. The AVM systems must share this spectrum with other users. Interference may come from industrial, scientific and medical devices and operation of government stations authorized in these bands. A table of these bands and other closely associated are shown below.

FCC BROADCAST FREQUENCIES

902 928	**General purpose mobile**
928-932	**Domestic Public Land** **Private Land** **Fixed Operational**
932-935	
935-941	**Private Land**
940-941	**General purpose mobile**
941-944	
944-960	**Auxiliary Broadcasting** **Domestic Public Fixed** **International Fixed** **Fixed Microwave** **Private operational**
960-1215	**Aviation**

TABLE: SELECT FCC BROADCAST FREQUENCIES

Proposed rule changes for the 902-928 MHz band would segregate location and AVI services on different frequencies, broaden the range of services that vendors could offer in this band, and possibly allow two or more location services to share the same portion of the band in the same market.

COMMON TECHNOLOGIES USED

So, as you examine the concept of frequency range, you might notice a number of recognizable names, like TV, AM & FM radio and the like. The concept of high frequency and low frequency RF/ID is placed on the chart so that you might get a perspective of where they fall relative to the spectrum shown. But there is more to the communication between the transponder and the reader device than the spectrum. Common technologies that are used include the Modulated Backscatter, the Surface Acoustic Wave, and Infrared. As these terms come up in discussions, there might be a need for more understanding of some of the principles behind each of the technologies.

MODULATED BACKSCATTER

Backscatter refers to the reflected energy from a source. The source sends a signal of radio waves toward the target, in this case a transponder. The transponder, using the energy absorbed from the incoming signal, contains circuitry that modulates the carrier with the data content of the transponder memory. This signal is sent back to the source. So, a modulated signal is sent back to the source. This is known as Modulated Backscatter.

SURFACE ACOUSTIC WAVE - (SAW)

Certain materials will oscillate if subjected to an illuminating signal. The oscillation causes an acoustic wave across the surface of the material. The material is known as piezoelectric material. A pattern of electrodes, placed by a photolithographic process, will detect the wave and generate a frequency that is then emitted to be returned to the illuminating antenna. Multiple electrodes are used to make a unique identification.

INFRARED

Infrared systems operate at the high end of the frequency range, above the microwave and radar signals. Infrared uses a higher frequency than the high frequency transponders. The signal has a short wavelength. Infrared tags are active. The signals will bounce off walls. Infrared is also used for motion detection.

FUNCTIONS

Fundamentally the more functions found in the system, the higher the cost for the system. This is obvious to most, but we continue to overlook this important fact as we evaluate systems.

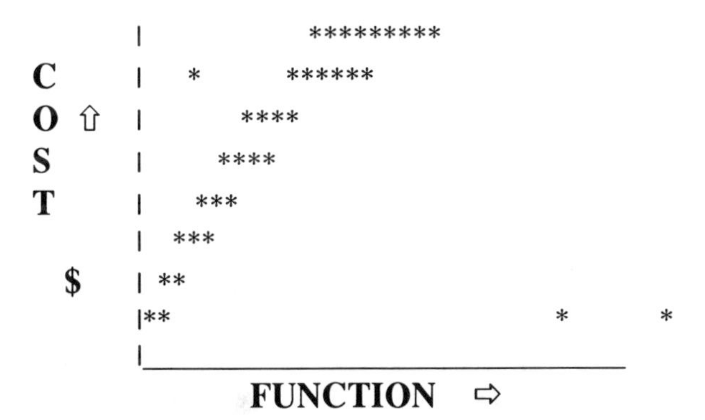

High function demands the proper respect of worth of the function. In general the more function the more the cost. There are some systems that have a rather low function and demand a high price because of the uniqueness of the offering. Others may have a lower cost because of pressures from competition and the price advantages of volume sales and other factors. As competition for this relatively new technology of RF/ID heats up there should be significant adjustment in the overall pricing. This chapter has function and components as the major thrust, so, onward to examine the functions required in the RF/ID systems.

Highlighted below are some of the key functions that come to the minds of end users as they articulate the need for radio frequency identification. Others may come to mind as the investigation continues, but if we are able to satisfy this list, there is a good chance that a successful system can be installed.

Functions Desired

1. Memory
   - Identification
   - Optional memory
   - External interface
2. Processor Control
3. Reader control buffer storage
4. Communications
5. Digital I/O Control
6. Analog I/O Control
7. Host System
8. System Health
9. Environmental Conditioning
10. Database
11. Pinpoint accuracy
12. Full area read
13. Security
14. Serviceability
15. Integration Options

MEMORY

For as long as computer systems have been in existence, there has always been a demand for more memory. There have been those applications that will say that there is sufficient memory in the basic identification that no other is required. While others maintain that as long as there is more memory available, then why not use it for the purpose of enhancing the functions that now can be automatically available. The truth is that the function of memory expansion is here in a competitive sense and will become a factor as long as the end user community takes control. To date, manufacturers have played a larger role in the specifications than have the users.

The memory of a transponder can serve many different end user functions. The size of the memory has little, or somewhat insignificant influence over the foot print of the transponder. As long as there is significant use for the memory, it is relatively free to the application. The caution is, of course, to insure that the memory transfer time and the speed of the tags on the objects do not cause operational slowdowns to conform to the tag rather than the opposite. Consider the fundamental uses for memory in the following table:

TRANSPONDER MEMORY FUNCTION TABLE

TYPE	*SIZE*	*FUNCTION*	*USE POTENTIAL*
on/off	1	Used as a single indicator - The transponder is either on or off, one or zero. Once changed from one state, function is over or destroyed.	Security systems
ID	128	Multiple bits, storage scheme for ID code and other items, like weight, type, owner. Limited read only.	Identify assets, boxes, pallets, equipment, autos.
ID Plus	0.5K to 8K +	Similar to ID above, but a variable memory for limited use for read and write.	ID function plus some dynamic area fuel level, status, etc.
Read/Write Design Limit		ID area plus an area to write variable data. Size is only limited to read/write time.	Toll, emergency, hazardous material, maintenance, route, etc.

TABLE: MEMORY FUNCTION TABLE

The early uses for memory were for a single bit to be on. Reader systems usually were placed at door ways like entrances to stores in a shopping mall. The function was to read the bit and turn on an audible alarm. This provided relatively low function for low cost to a user community that could see significant value for a security system for the protection of high cost items. Identification of specific items caused a user demand for specific items. The memory sizes increased and coding schemes were designed to not only identify the asset, but had the ability to hold special codes and descriptions. As read/write systems were introduced, the size of the memory began to increase to where there are rather large storage capabilities available from some manufacturers. Certainly, thousands of characters can be represented and before too long, megabytes of data will be stored on transponders.

Not only is the primary storage of a transponder expanding but what might be called secondary storage is also possible. First of all, the concept of having an external input source for part of the memory is available in a function usually referred to as a dynamic function. A tethered connection to the transponder or tag provides an interface to the storage. Other connections involve even more elaborate capabilities, like using SMART cards or credit cards as a source of data storage functions.

What is certain is that whatever the memory size, it is likely to be in high demand as the applications that are now possible stretch the size limitations of today.

MEMORY CONCEPT

FIXED	**SET IN THE PLANTM OR ENTERED VIA USER THROUGH AN ENCODER**
OPTIONAL	**FEATURES AVAILABLE**
DYNAMIC......................................	**TETHERED CONNECTION GAGE, METERS, SENSORS**
READ/WRITE	**MEMORY TO BE READ AND WRITTEN TO BY READER STATIONS.**

TABLE: MEMORY CONCEPT

PROCESSOR CONTROL

Processor control is used to manage the functions contained on the tag and the control unit functions of the reader. Coding schemes and command control sequences are needed to insure reliable and accurate data. Processors have typically been increasing in speed, and function, while decreasing in foot print. As the transponder functions change over time, the significant application demands will be placed on the processors. It will be a requirement to read all generations of identification tags, so the reader systems will require logic associated with all levels of the population in use. The processor speed is important to the overall systems performance.

However, while processor speed gets into the decision cycle an analysis of performance needed might assist the user in the choice. The slower processors tend to be less expensive, as they have been made obsolete by successor systems. When the application matches, there can be significant benefit for the user, since system performance is related to other functions in the system, where more critical improvements can be made.

READER CONTROL BUFFER STORAGE

As readers perform the read and validation function of a tag- read, they store the event in buffer memory usually along with other data like time stamp information. This function stores the data prior to sending it to a host system. The storage size should be large enough to hold sufficient data to cover the peak outage between the reader and the host system.

COMMUNICATIONS

Standard capabilities to communicate from and to the reader, and the host is required. Depending on network availability, dial up service, dial out service and leased line capability may be required. Users have requested data radio communications as well, for applications to reach a remotely situated reader where power and telephone service were not available. The better the communications options, the better overall installation and more serviceable the system will be.

DIGITAL I/O CONTROL

The applications can be more elegant when the function of the system does more than merely identify the transponder. Other data collection devices can provide inputs in the capture process. Output signals can control devices such as gates and cameras that can be activated from I/O control ports. These ports are offered via reader equipment in a stand alone configuration and are a part of many standard computer systems in terms of I/O cards.

ANALOG I/O CONTROL

In the case of Digital I/O, the application may require the use of Analog I/O. These can provide significant functions for the end user.

HOST SYSTEM

The identification system uses the processor control to read the tag automatically and collect the necessary inputs from devices on the I/O Controls. The data is then sent to one or more Host Systems depending on the application or set of applications. The Host System has two primary functions.

1. Acquire, store and process the identification data.
2. Provide reader service functions to a network.

The reader provides basic collection while the host can provide more sophisticated processes on the data. Analysis reports can be generated from the stored data. Messages can be sent from the host to other systems. Reader service functions are also needed.

To assist in controlling one or many readers, it is important to be able to obtain status information from remotely situated reader stations. The requirement is to access onc or all of the readers, so specific requests must be possible. Reader diagnostic functions can be sent to the reader from the host to assist in problem diagnostics and testing functions.

SYSTEM HEALTH

Overall system health can be determined with a collection of evaluations for each reader and a history of individual reader system performance and maintenance actions for the reader. This function can be run on demand and can also be part of an overall preventive maintenance operation.

ENVIRONMENTAL CONDITIONING

While some RFID applications are situated in environmentally controlled environments, the readers are likely to be in rather harsh conditions. To accommodate this practice, a set of environmental control features should be available in the packaging options; heaters for cold conditions, and air flow and air conditioning for hot environments. Surge protection and lightening arresters need to be considered. Sometimes the tags need special features to handle the environment as well.

DATABASE

To insure availability of data and ease of access, the system should be organized by using a database manager. The collected data can be housed under a database scheme and any status records can be kept under control of the database.

TAGS - TRANSPONDERS

READER EQUIPMENT

PINPOINT ACCURACY

The application systems will require that the reader has the capability to read the transponder next in line. This capability assists in applications where association of a transponder will correspond to other collection activities. So identification is for a specific ID. Here the transponder passes through a choke point and when it does a read is required. Some applications require that the reader pass by the choke point where a tag is placed.

FULL AREA READ

Certain applications will need to identify all transponders that are in an area or are passing through an area. In this case, it is important to identify every transponder in the field of interest.

SECURITY

Security functions are required to insure data integrity and authorized use. The concept here is that it may not be desirous to have just anyone read a transponder. In the case of read/ write functions, you may desire to control who writes to the tag. As data moves from location to location as part of a transponder, you may want to have a system where control codes and encryption techniques and options are available for your use. Therefore, it is important to control who is going to have access to the data on a transponder.

In addition, consideration should be given to the ability to copy the transponder. The easier it is to counterfeit the tag the less secure the system is.

The system should be reasonably able to handle jamming situations. In some cases the system should at least alert the controlling location that there is a possibility of a jam in progress.

SERVICEABILITY

The system should be designed to provide for ease of repair and ease of modification. Care should be taken to understand the manufacturer's design and maintenance philosophy. Check to see if field replaceable parts and subsystems are available. Another check should be on the manufacturer's product releases. Are the parts interchangeable or mutually exclusive?

INTEGRATION OPTIONS

Of course, all of these options and features must fit into the environment easily. If the system can be installed using the currently installed equipment with minor modifications, than the current investment is protected.

SUMMARY

The basic components of the RF/ID system, transponder, antenna, reader and encoder, offer the fundamental building blocks for the system. The functional requirements however imply a much more sophisticated system.

CHAPTER III

RF/ID INTRODUCTION

Radio Frequency Identification has been used for multiple applications, and the use is expanding. RF/ID is a contactless identification, and there are a number of advantages and capabilities that will revolutionize the identification industry. There are different tag types available. These exciting functions provide significant opportunity for re-engineering the processes.

An alarm goes off as I leave the shoe store and I hurry back to the register. "Sorry!" the clerk says to me. She then takes my new shoes out of the box, runs them over a flat device on the counter. This deactivates the transponder hidden inside a shoe, and she sends me on my way. North American railroad cars should have two transponders, one on each side of the car for identification purposes by the end of 1994. European rails will install transponders in the road bed for high speed communications. Truckers pull up to a fuel station and a device identifies the rig. Automobiles drive through lanes without stopping and tolls are paid through the use of the RF/ID system. Assembly lines have RF/ID tags assisting the production and automation of the plant. People are identified as they pass through the entrance to their buildings without digging into their wallets and purses. Radio Frequency Identification (RF/ID) is hot, hot, hot! There are more and more users of RF/ID systems. Why is this RF/ID business catching on, and what is it all about?

Information needs continue to grow. Accuracy and timeliness of data provides the fundamental makeup of information in today's competitive environment. For years there has been this need to improve the accuracy of the data. Systems were put into place with manual key entry. Verification systems would assist the users in providing high quality data. But, the overall process for providing accuracy of the data was too costly for many applications, and thus, there was a reluctance to install these manually inaccurate processes. When the data was thought to be accurate, we relied on the manual process to make the data available to the systems that would use it. Thus, while using computer assisted data entry and verification systems for increased accuracy, we relied on a manual process to get the data to our systems. This caused us a timeliness problem. We batch the data to get it to our systems.

The concept of the Radio Frequency Identification system is a simple, yet very powerful one. Transponders contain the identification number or code permanently affixed in the memory. A reader system reads the identification in the transponder as they come in close proximity to each other. When the antenna of the reader receives the signal from the antenna of the transponder then the coding schemes can decode the information to insure accurate capture of data. Typically, the antenna is placed in a fixed location, like a doorway or gate, and the transponder is read as it passes by this fixed point location. The system then, because it is a computer with a mission to deliver the data, can do so with timeliness and repeatability.

There are three particular features of radio frequency identification that come out in discussions with the user community. They affect the information needs of the end users and the flexibility of uses required for automatic capture of identification data. The three features are:

1. Operates in a relatively harsh environment
2. Long distance between transponder and reader/antenna
3. Larger storage capacity than traditional identification.

HARSH ENVIRONMENT

Today the ability to identify things in a lights-out situation, or in an environment of rain, fog or snow, produces the advantage many users have requested for a long time. The dirt and grime of the plant sites and general roadways do not seriously impact the ability to read the transponder. Railroads and intermodal carriers tried special bar codes in the 70's but the environment rendered the bar codes ineffective.

The harshness of the rain, fog, dirt, and in general, the elements are considered to be the downfall of traditional bar codes. New views of harshness come to the forefront when you use computer circuits to produce the radio frequency identification system. Interfering signals from electrical currents and magnetic fields, shock and vibration of the transponders affixed to equipment and packages, and quick changes in temperatures will impede transponder performance. The manufacturers have, however, designed the transponder to live in the environment and to overcome these impediments to provide a highly reliable and available system.

READ DISTANCE

The read distance varies with the type of transponder, and the characteristics built in by the manufacturer. Generally, a transponder can be read from as close as a few millimeters to as far away as several hundred feet. This provides a window of read opportunity that can be controlled through the use of special antennas.

STORAGE CAPACITY

Since the transponder is primarily a computer chip, the storage can be of a capacity to handle the wants and needs of most users. So, it is not only an identification but also a source for information. This storage can be permanent or variable. The storage can be used for many different applications. Maintenance records, contents history, hazardous material records, inspection records and a fundamental interchange capability are a few applications, and there are many more options.

The idea of using storage media and a basic identification label for historical record keeping has significant promise. Standards are beginning to surface and are being refined to assist the user community with the fundamental answers to their wants and needs.

CONCERNS

For those primary advantages there are two primary concerns that enter into the equation. Of course the cost of a transponder is certainly significantly higher than the typical cost of a bar code. Several dollars to seventy dollars might be a typical range for an RF/ID tag. The cost for specialty tags might be even higher than seventy dollars. As the tags become more competitive and the use increases, you can expect a dramatic reduction in cost. There is a pent up demand for identification systems that can be delivered for less than one dollar. The size of the transponder is the second concern. While the sizes of RF/ID tags vary, they tend to be thick and bulky when compared to bar codes, or even magnetic stripes for that matter. To retrofit the tags into the application may cause additional start up costs until the use is fully integrated.

FIXED POINT READER EQUIPMENT

The idea is that while the transponder or the reader equipment travels about, the other is stationed at a fixed location. Here are a couple of examples. Transponders affixed to rail cars can be read by fixed location readers at track side. As the train passes by the fixed reader, the identification is made. Readers placed on buses can read transponders placed at bus stops along the route. As the bus with a reader passes by a transponder fixed to a roadside post, the identification is made.

The transponder is read through the use of an antenna and a reader device. A reader and antenna might also be called an interrogator. Signals are sent to the transponder, and the transponder reflects a signal back to the interrogator with the identification codes. The reader system decodes the identification of the transponder.

Some systems are designed to capture one transponder at a time. This would be, as an example, the transponder with the strongest signal being reflected. This concept is used when it is necessary to identify a specific transponder. Other systems have options designed to read any transponder in view of the antenna. The use of Time Division Multiple Access, (TDMA), provides the ability to read multiple transponders traveling across an antenna path, in relatively the same time or remaining in the view of the antenna at the same time.

TRANSPONDER TECHNOLOGY

So, the radio frequency tags with the identification codes provide the means

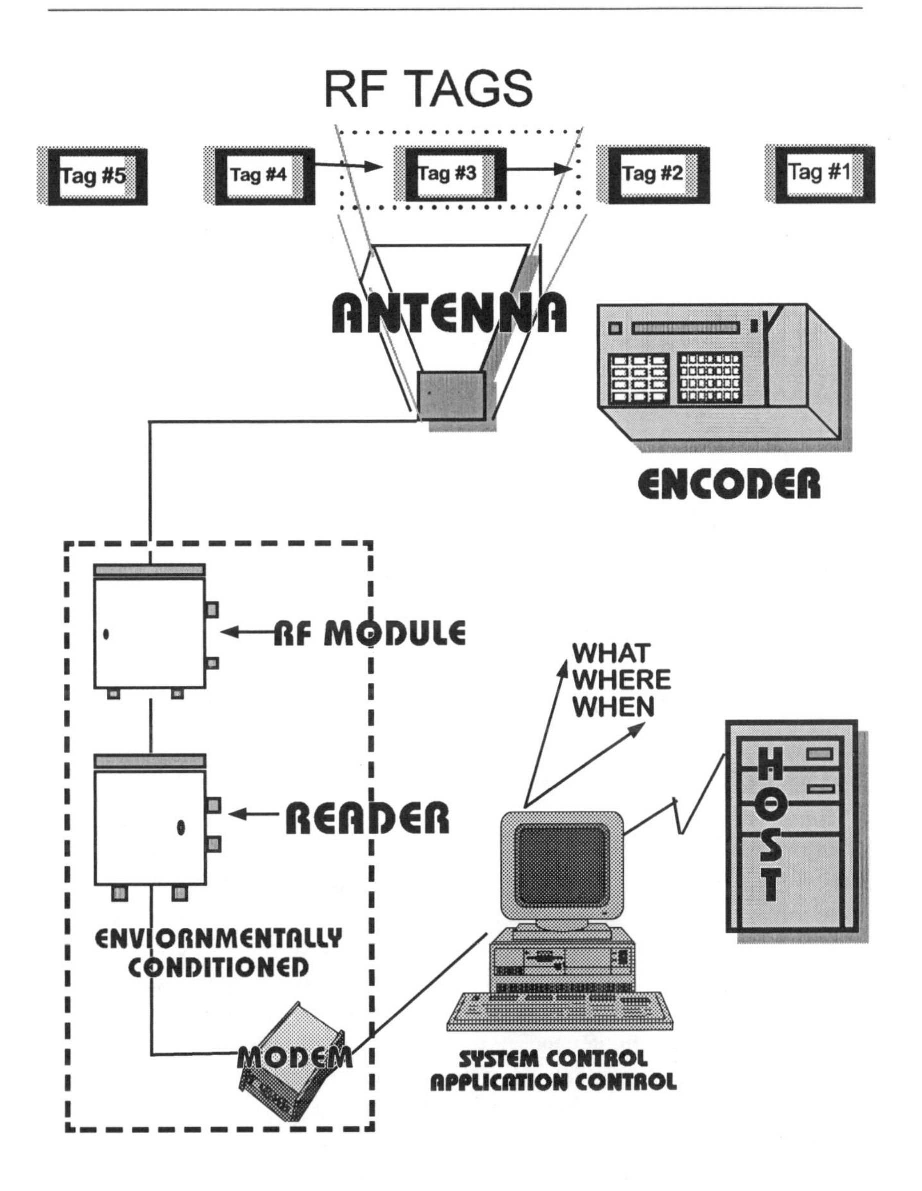

RADIO FREQUENCY EQUIPMENT

RF SYSTEM DESIGN

for automatic equipment identification. This is exciting, but the forward thinker gets even more excited about the possibilities of expanding the capabilities of the tag. In other words, the tag technology is bringing more and more function to the table. There are several types of tags emerging. At the risk of causing some confusion about tags and tag types, it is worth trying to understand the possibilities.

A. Read-only tags.
B. Dynamic Tags.
C. Read / Write Tags.

Read-Only Tags:

The read-only tags are the ones that contain the identification and perhaps a little more information. The data is permanent and can be read by a reader.

Read-only tags have identification as the primary function. Before the tag is placed on an object, the tag has an identification encoded into the memory. This encoding process can be enacted at the place of manufacture, or in the field through a special encoding device. Once the tag is encoded, it will hold this identification permanently. It is primarily intended to keep the unique identification. It is possible with some manufacturer's devices to change a tag. The process is not one for routine modification. When placed in use, the tags will be read by a reader system.

Dynamic Tags:

These tags contain the identification information similar to the read-only, and also, changing data, as might be available from a fuel gauge or thermometer. A tethered connection from the device to the tag serves as the data channel.

The possibility of automatically identifying a tag as it passes by strategically placed readers is exciting. The possibility of getting variable data in the same fashion is revolutionary. The concept of the dynamic tag is gaining strength. The concept of a fixed portion and a variable portion of the tag is available for users. The fixed portion can be thought of as similar to the data contained in a read-only tag. The variable portion can contain many things. Concepts like fuel levels in rental cars could be read so that improved customer service on returns is possible. Maintenance information could be stored in the dynamic area. Shock and vibration data for transport vehicles can monitor the cargo as it travels across the land. You can bet that certain shippers would love that kind of service information! Think about the hazardous material records that could travel with the car. Some manufacturers have built-in credit card and smart card interfaces for use in toll collection. The applications seem limitless.

Read/Write Tags:

If the concept of the dynamic tag is to have the collection devices and meters feeding the dynamic portion of a tag, the concept of read/write is for the interrogator to read a fixed and variable portion and write directly to a variable portion. The read/write concept features a fixed identification portion of a radio frequency transponder. There is a variable portion as well. The user can not only read the fixed and variable portion, but can also write to the variable portion. The read/write tag has many applications. Some of the applications might be similar to that found in the dynamic tag but many will be very different. Considerations of memory size and other features start to surface.

If the memory of a read/write tag were sufficient, the description of the contents of the rail car, container or truck could be written on it. The maintenance records of the car and inspection data could be kept on it. It could be possible to have operational detectors, like railroad "hot box" detectors, write recorded readings on it. Bills of lading and hazardous material information could be written on it. Yes, there are many applications that are possible.

AWAKE AND TELL TAGS

Some tags are being used that monitor satisfactory situations and do not transmit, even with reader interrogation, until the out-of-specification situation or threshold is met. The tag is quietly monitoring the situation and then, when the condition or set of conditions are met, the tag transmits the condition to the reader. This kind of tag can offer the value of having less data to process. If the tag always responded, even under normal or satisfactory situations, the data would need to be processed.

RE-ENGINEERING

The particular uses of the AEI systems should be viewed from the standpoint of user requirements, and the potential benefits to be derived. Here is a great opportunity to provide the real value of quality control within your organization. The implementation of AEI can produce some very significant benefits. The benefits come from improved processes and controlled implementation of change. The good news is that AEI produces greater accuracy of identification data on an automatic basis. The bad news is that it is automatic from the point of view of current processes.

What this means is that an automatic system will produce automatic answers. Your current system may require only one answer for a process. If two or three answers come to that same system, because it is automatic, then there may be big problems. A re-engineering is required to insure a quality implementation.

The re-engineering involves a significant effort to consider the options available for gaining the biggest advantage for the organization. There are three cases that might help you in your efforts with AEI systems.

1. Simple Applications - Stand Alone
2. Integrate Into Existing Applications
3. New Strategic Applications

SIMPLE APPLICATIONS - STAND ALONE

This approach determines an application that is relatively simple. The application can provide benefits to the organization, but is not so complex that a major investment in system development and programming systems are required. One example of a simple application was a gate system that read the tag and immediately printed the tag ID with a date and a time stamp. Operations management could go about other business, then return to the gate house to see what had come in and out of the gate. As the operations management were observed, you could see their delight as they discovered a particular trailer had entered the yard. Staying with the gate for a minute, one customer had the various vendors and service crews tag their maintenance vehicles. Again, the approach was to print the entry and exit data as a printed log, documenting service commitments. There are other examples.

Parking lot control is another application that can be implemented without much investment and still provide benefits. One customer placed tags on shuttle busses to determine service levels for employee pickups. There are also closed loop applications that would permit part of the fleet to be tagged for identification at a limited number of locations.

INTEGRATE INTO EXISTING SYSTEMS

This particular step can produce significant value if there is a problem with the accuracy and timeliness of data entry and if the current system uses an identification code. It may be possible to reduce the amount of effort needed to key enter the data with the automatic ID system. The following is a conceptual example.

This example shows how an existing system, SYSTEM A, has five data entry clerks to feed the system. Let us say the operators one through five would be given batches of data to be key entered, validated, reconciled and so forth. They would perform these duties and control the process. Compare this scenario to that of the AEI system.

Here we see operators A, B, and C. This is to imply that a new procedure found a savings of two operators, five minus three. Keep in mind that this is an illustration and not meant to be an indication of average or normal savings. Your situation might be totally different. But notice that to gain the efficiencies, a modified system is employed. Note the NEW SYSTEM diagram and the location marked MODIFICATION. The Modification includes procedures to assist the operator in validating and controlling the process.

MANUAL KEY ENTRY

operator 1------------ **operator 2------------** **operator 3------------** **operator 4------------** **operator 5------------**	-------------------------------- -------------------------------- -------------------------------- -------------------------------- ----------------------	**SYSTEM A**

NEW SYSTEM- AEI

operator A------------ **operator B------------** **operator C------------**	**modification**	**SYSTEM A**

INSTALLING AN AEI SYSTEM MAY REQUIRE SOME NEW OR MODIFIED CODE

FIGURE: CONCEPT INTEGRATE INTO EXISTING SYSTEM

Assuming the same workload, it would mean that with automatic identification and a modified code, three people could do the work of five.

NEW STRATEGIC APPLICATIONS

This is an area that is most exciting. Here is where one can be as creative as possible, and effect change in processes that can revolutionize how the organization will operate. As an example, the following figure is provided to show what might be possible. Since I have used rental cars in my travels I will use a process for rental car returns, and make a few assumptions of system richness.

AEI RENTAL CAR RETURN

OLD WAY	NEW WAY
1. Drive into a gate 2. Park in slot 3. Write down fuel level 4. Write down mileage 5. Go to counter 6. Wait in line 7. Get receipt 8. Go to bus 9. To airport	1. Drive into a gate 2. Park in slot 3. Pick up receipt on bus 4. To airport

FIGURE: EXAMPLE NEW PROCESS RENTAL CAR RETURN

SUMMARY

Radio frequency identification systems are providing accurate and timely data collection. The identification can be made from long distances in rather harsh environments and the storage capacity can be significantly large. Read-only, dynamic and read/write systems of radio frequency identification applications are possible now to help solve strategic problems. The dynamic tag is gaining strength, and standards are beginning to be formulated. The read/write technology is being refined in the toll marketplace, and other concepts are now being developed. The technology is far along enough to use it to improve customer service, operations, or to streamline your processes. Just imagine the possibilities!

CHAPTER IV

THE SPECTRUM OF HARSHNESS

Identification systems must survive the harsh environment. There is an entire spectrum of harshness that is ready to force its wrath on any identification system that dares to venture into the elements.

The term, harsh environment, immediately brings to mind cold blowing winds carrying the ugly, bone freezing cold across the land, and piling snow and ice on my driveway. Of course, a harsh environment for an identification medium has a much broader meaning. Simple rain to hard driving rain storms, to the peak of the wrath of a hurricane or tornado fall in the spectrum. The acid rain of chemicals in the environment, the heat of the burning sun, the changing extremes of hot and cold, and dry and wet, qualify as well. Then there is the notion of dirt and dust. The qualities of vibration and shock are in there some place as well. Now and again the environment might have waves of magnetic fields, caused by rivers of electric currents, or radio frequencies, dancing and jamming about. Throughout the spectrum of harshness, there is a need to identify things as they move about the environment. The medium that carries the identification must survive in the environment in which it is placed.

CONTROLLED ENVIRONMENT

For a long time, we created controlled environments for identification. We kept things in cartons, cartons in boxes and boxes on pallets. We could label the things, the cartons and boxes, and the pallets we could count. (More on pallets later.) While the things are on the assembly line we control how they are handled. We put the assembly lines under the cover of buildings, and then control the lighting, the humidity and the temperature. We put a paper bar code on everything we possibly could. Bar codes here and there and everywhere! We even placed bar codes and other special coding on our paperwork and readers at our desks. The more we used them, the more we found that a controlled environment was best. Harshness was delivered by mishandling the bar code in some way. We tore the bar code, or we smudged the bar code. In general, we destroyed the bar code. We spilled coffee on the codes. So we developed better bar codes. Longer, wider bar codes made them harder to destroy and increased the probability for a successful read. We found this capability so rewarding, we began to expand the use beyond the controlled environments.

SEMI-HARSHNESS

Bar codes were placed in an environment like a factory floor where it was determined that the environment had more dirt and grease and oil. Magnetic stripes were placed in cards and coated to protect them from the corrosive atmosphere. The densities of the metal oxides on the magnetic striped cards provided more data and the coatings maintained the higher reliability and repeatability of the read. We relaxed our usage requirements for the controlled environment. As this came about, the creative engineering community had to protect the medium even more. New bar code techniques and protection schemes increased the usage beyond the traditional office environments. The harsh environments of factory floors and loading docks and bills of lading that traveled with drivers caused us to re-engineer the medium. The technology continued to develop in such a way that bar codes could be read from copies and even crumpled paper. The magnetic stripes could be read reliably, and reused with more data than could be found on bar codes. We could even carry the magnetic stripe with multiple others in the form of credit cards in our wallets. Sit on them, rub them, use them a lot and use them we did. We continued to spill our coffee on them, pour the toxins of the soda fountains on them, but they worked for us. We kept them in the cold, in the heat, and, we kept them in our cars and trucks a lot.

At this point, we used bar codes and magnetic stripes at points of contact or near contact. We advanced the medium another step as we found that it was important to get the identification from distances beyond contact. We wanted more distance. Distance would provide additional flexibility in the use of the identification medium. The medium became more reflective. Multi-color was used for specialized applications. Larger code sizes helped us read at distances never before imagined. We could then have application uses that were several feet away from the object. We began to see store checkout systems that used holography to read the labels. We saw a new radio technology that was used to help secure the items in the stores. These had limited distances associated with the label, but they were sufficient to place reader devices in doorways and entrance ways in stores and malls. While the radio devices were more expensive, they were reusable and provided a long distance function.

HARSH ENVIRONMENTS

We began to like the notion of the radio device. There was the advantage of being able to read the device in a variable space. Most of us remember the early transponders, placed on wild animals that helped to follow the animals migration patterns. It seemed that the high function of the radio transponders was an advantage. There was more memory possible, more read distance available, and the transponder could be read through fog, rain, snow and mild coverings of dirt. Computers had advanced to the point that, now, a computer chip had the miniaturization required to provide an opportunity to package the chip to survive the environment. So, now, the pallet could be identified as it traveled through the veritable free space of the warehouse and loading docks. Trucks and trailers, and chassis could be identified on a reliable basis.

The idea of producing a card with enough electronics to provide high function and enough memory to provide additional new applications, had operational management creating new justifications. We covered the computer card like we covered the bar code. We packaged the computer card to protect it from the harsh environments of not only nature, but the electrical and magnetic interferences found in our modern environment. I remember viewing an antenna that was used to read the transponder. It was covered with snow and ice. I mean, it was buried in snow with nothing but a tip of the antenna showing. There was a radio station next door with mega watts of power, trying to interfere with the antenna. But, the transponders were read.

We have improved bar codes and magnetic stripes, as well as the new radio frequency transponders. So as harshness calls the freezing cold of snow to my driveway, I wonder what will they think of next?

With that as a background, it is important to understand just what harsh environments are, and to examine what is important in the various categories in the spectrum of harshness. A look at the Table, SPECTRUM OF HARSHNESS will show some of the more important aspects of the spectrum.

Harsh means sharp and unpleasant, cruel and severe. This connotation gets us to focus on the extremes of the environment in which the system is to operate. Severe cold can stop the function as can severe heat. Power surges from lightning can burn out electrical circuits and lightening is considered extreme. Another concern is the transition from one state to another. If a cold silicone chip is heated at a fast rate it tends to explode. A sharp point can cut and destroy. A magnet can wipe out the contents of a diskette or magnetic medium. These are harsh environments.

The Spectrum of Harshness is offered as a concept, to help us solidify the notion that the identification system must survive in an environment that has not been traditionally thought of as safe for accurate and repeatable identification. Quite the contrary, the harsh environment was always thought of as a forbidden place for computer systems, and this would apply to data collection systems as well.

What is harsh to one identification system might not be harsh to other identification systems. A simple example or two might help here. A bar code on a paper label can be read by a laser scanner bar code reading system. Paint the bar code and the bar code cannot be read by the same scanner. Paint a low frequency identification transponder and the reader can still read the identification even though the outside has been painted. Another example might be to place a transponder in a purse inside a car. There are low frequency identification systems that will read the ID in the purse inside the car, but high frequency systems need line of sight to read the ID, so this type of reader system would not read the high frequency identification transponder in the purse inside the car. Secondary influences might also impact harsh environments. A bar code could be read in a line of sight fashion, but a high frequency transponder could provide inaccurate results if another transponder was being reflected in the path. So in the first example, paint was harsh to the

ENVIRONMENTAL CONDITIONS
ENVIRONMENTAL CONDITIONS:
DIRT, ICE, SNOW, AND
DRIVING RAIN, INTERFERE
WITH THE READ PROCESS
OPTIC

bar code label but not to the radio frequency transponder. Likewise, line of sight was required to read the high frequency transponder in the second case. And finally, a bar code label was read in the third case because the environment caused a reflection.

SPECTRUM OF HARSHNESS

Water	Tornado	Acid	X-ray	Break	Shock	Heat
Rain	Hurricane	Chemical	Infra-red			
Fog	Storm	Grease	Microwave	Hole	Bump	Hot
Humid	Wind	Dirt	TV	Crack	Nudge	Warm
Slush			FM	Scratch	Shake	Cool
Snow	Breeze	Pollen	AM	Dent	Rattle	Cold
Ice			Inductive	Check	Vibrate	
Dry	Calm	Clean	DC	Smooth	Still	Freeze

TABLE: SPECTRUM OF HARSHNESS

The environment in which the identification system is to survive is critical to the success factor that will result. This applies to both the label medium and the reader system. Both must survive the environment. If the label survives but the reader system does not, then significant system failures will result.

Some of the factors included in the SPECTRUM OF HARSH are as follows:

A. Humidity
B. Wind
C. Chemical
D. Energy
E. Physical
F. Stress
G. Temperature

HUMIDITY

The harshness of humidity has long plagued the computer environment. For years computer rooms could be thought of as too high or too low in terms of humidity. Special devices were placed in the computer room to measure humidity and other devices were placed there to take corrective action. Static electricity could cause significant damage to storage devices and to the logic of programs in the computer's memory. There is a subtle relationship between humidity and the operation of the computer device but as far as the identification medium is concerned, there needs to be a validation of operation in the humidity of the environment.

The relationship of humidity to the label present different problems based on the type of label. A bar code label printed in water soluble ink will not survive

DANGER
HIGH
VOLTAGE
DANGER

a rain. For that matter, neither will a radio frequency transponder in the long term, unless there are special conditions. The bar code label would dissolve and the transponder would corrode over time. Actually, we can simply apply a coating to each type of label to protect the information contained in it. We can protect the label and increase the probability of a successful read. Let's solve this problem by using ink that survives the application of water. What about being submerged in water?

Depending on the type of label being used, one condition is that a label should survive being submerged in water to be read at a future time. The other is an ability to read the label while it is submerged in the water. There are a few transponders that can be read while submerged in a few inches of water, but there are not any that are known that can be read from large depths. This also applies to bar codes.

Typically when the labels are presented to readers in conditions such as rain and snow, the radio frequency transponder has a better chance than the bar code. This is true even in heavy down pours or blizzards.

What happens if the water freezes and ice forms in a cake above the label? Certainly this is thought of as harsh. Some labels have been improved so that an inch or so of ice can cover the label. Again, the low frequency label has an edge here.

WIND

For the most part, wind does not effect bar code or radio frequency identification labels. The key is that the wind usually brings other harsh elements with it. Dust, snow, and rain are some examples. Wind has a major effect on the reader systems and subsystems. Because of the sensitive electronics contained in the readers and other equipment, engineers have housed the equipment in special packaging or enclosures. In fact, the high function transponders also have special packaging. The enclosures used protect the electronics from the harsh environment presented. In other words, the container is designed so that the wind does not cause the other elements of the spectrum of harshness to enter the container. The wind cannot drive the rain through the container or throw objects into it, causing physical damage.

CHEMICAL

An environment of chemical exposure is at the top of the list for its negative impacts on the identification system and the subsystems used to control them. Acid can burn off the label, and cut the protective coating on the transponder. In general, the transponder may win out over the paper label, but neither is given a good chance in the long term. Depending on the chemical and the qualities of the chemical symptoms, the reflective qualities may be altered, or the introduction of electrical currents or alternative currents are possible.

These tend to render the system useless.

In fact, it is considered better when these conditions cause permanent failure. Far too often, there is the dreaded intermittent failure that results. The intermittent failure often cause us to reach an erroneous conclusion about the system or subsystem and prevents or at least delays progress.

ENERGY

Bar codes are not effected by the introduction of electrical currents or radio waves into the environment. Radio frequency transponders on the other hand can be sensitive to this environment. The reader systems of both can be affected in a negative way by this type of contamination. The problem is that you cannot see the contamination, and often, it is not easy to determine the source of the problem.

In some cases, the energy may come from external sources such as adjoining property. Radio stations, electric power plants, and radar stations are examples. In other cases, the source of a disruption may come from objects as they pass by the location. A locomotive pulling a train has steel wheels that ride on steel rails and can generate electrical currents and magnetic fields. In still another example, the energy might be reflected from metal containers, as they travel here and there within the location of the reader site. So, an environment might be good when the parking lot is empty, but not so good when the parking lot is full. It might not be good when one particular parking bay is occupied.

The point is that certain energy levels can render the identification system inoperable. The source of the interference might be from a permanent source or might be from a more dynamic or variable condition.

PHYSICAL

Identification systems are being used in applications where it becomes more important to operate even with some amount of physical damage. Bar codes can be scratched or torn and still provide an accurate identification. An RF transponder can survive, even when the casing is cracked or a hole is poked through it. In both cases, of course, the damage could be severe enough to cause system failure. There have been cases documented that have featured transponders being shot and still providing an accurate identification even though there was a large hole in the case. There must be some resilience for the system to be dependable. Physical damage should not always render the system unusable, but there are other stress points that can effect operation.

STRESS

Stress in this section is in the context of physical stress. The concern is for what is happening to the label or the reader throughout its life. Two important elements are vibration and shock. The label can be mounted on just about everything these days, a box, a pallet or a piece of equipment. If the label is subjected to constant or long term vibration or severe shock, will it survive? A quality testing procedure would include vibration and shock tests to meet the specification of the user. If a simple tap shocks the system and breaks it, or if the mere act of transporting the device to market breaks it, then there would not be much use for it. What we are looking for is reliable operation over a useful lifetime, despite repeated episodes of vibration and shock.

There are levels of tests that can be applied as there are levels of engineering design. The more rigorous the test and precise the design for this environment, then the more costly the system might be. There are vibration and shock tests that apply to the office environment and there are tests for a more rugged and military environments.

As an example, there are drop tests that drop the device in a very controlled way to see if the device will operate after being dropped. The device can be dropped from a height, let us say, of four feet. The drop test could include a flat drop, and corner drops. So the test engineer would determine if the device would operate after being dropped from a height of four feet onto a cement floor with the device landing flat on its face or back side in the first case and on a specific corner in the second case. Usually all corners are tested to see if there is any vulnerability to any side of the device.

TEMPERATURE

Just about everything we do involves some type of temperature control and this consideration is indeed a necessity for identification systems as well. An environment that is too hot or too cold can cause the system to fail. There are differences in operating conditions and in storage conditions.

Storage conditions are generally more extreme than the operating environment. If the object reaches its maximum temperature extreme during storage, the object would have to be brought to within the operating temperature range prior to operating or permanent damage could result.

Categories of harshness have been identified for temperature ranges. These categories help us determine how a specification might meet the needs of the expected use. Here is an example of categories.

TEMPERATURE CATEGORIES

1.	Office	17 - 23
2.	Commercial	0 - 55
3.	Industrial	0 - 70
4.	Extended	-40 - 85
5.	MIL Rugged	-55 - 90

The idea is that as the requirement for more extremes in temperature increases, so do the design requirements for the device. The military specification shown provides for the harshest environment.

The notion of operating temperature is one that design engineers have to pay significant attention to. The particular problem is in controlling how the device is brought to the operating temperature prior to operating. Computer chips tend to explode if they are colder than the operating temperature, and an electrical current is forced through the chip. Special start up circuits would need to be implemented to cause heaters to activate bringing the device to the operating temperature.

Of course, there are designs that attempt to keep the critical circuits at operating temperature. Heaters are used for the cold environments and fans are used to cool in the hot environments.

SUMMARY

The SPECTRUM OF HARSHNESS covers a broad range of considerations for a well designed identification system. Some of the aspects within the spectrum can cause permanent damage and others temporary damage. Some within the spectrum are immediate, in terms of how it affects the system and others are more long term. Consideration must be given to the "Spectrum of Harshness".

CHAPTER V

STANDARDS

Standards are required to gain the most from this technology. There are so many opportunities for the RF/ID systems. The interactions within the user groups is such that the system will benefit if standards are set. The caution is to allow the continuing evolution to take place to accommodate new options.

A standard is anything taken as a basis of comparison. It is a model to be followed. Standards have been a cornerstone to the computer revolution and the identification community. Without standards the user community would have significant troubles in communicating with their constituents, gaining significant productivity from common capabilities, or having a point of comparison reflecting the views of the experts. Standards have long been thought of as having both good and bad aspects. A good aspect is that once the standard is set there is a benchmark for the suppliers of the world to follow. A bad aspect is that standards can sometimes prevent or stall new concepts from entering the market. Generally, politics surround the formation of a standard. There is also a significant amount of technical engineering support.

Fortunately, there are standards that can apply in the formation of a new integrated system. These standards might reflect the environmental conditions needed for the survival of the equipment, or the safety features required to prevent bodily injury. There are standards compliance laboratories, like Underwriters Laboratory, to test and insure compliance. We often look for the UL approval seal. There are packaging standards that reflect the needed levels of protection against the elements. The military and space programs have produced both generic and specific standards for our use. There are interface standards for equipment and power. There are standards that are backed by industry organizations and those backed by international organizations.

Standards organizations that have specifically addressed the radio frequency identification environment are listed below. These organizations have levels of approval for standards. It is always a good idea to contact them for the latest documentation.

AAR Association of American Railroads Standard for Automatic Equipment Identification was adopted in 1992.

ANSI American National Standards Institute approved a standard for Freight Containers automatic identification in 1990.

ATA American Trucking Association specified a standard for Automatic Equipment Identification in 1990.

IATA International Air Transport Association has recommended Practice 1640 Use of Radio Frequency Technology for Automatic Identification of Unit Load Devices.

ISO The International Organization for Standardization produced ISO 10374, Freight Containers - Automatic Identification in 1991.

These organizations have applied their standards with a set of common concerns for the health and reliability of the identification system. If one reviews the standards, there are significant references to various military specifications that reflect a rigor proven in the past to be significant to this type of application. In addition, care was taken to specify the best location for placing the equipment, such as the radio frequency transponder, how to orient it, how to affix it to the equipment and so on. The contents of the data bits were also described for the user. Here is a list of some of the common items of concern:

A. Reliability
B. Accuracy
C. Tag Life
D. Speed
E. Temperature
F. Frequency
G. Tag Life
H. Tag Position
I. Data Content
J. Distance

RELIABILITY

Reliability has do with how much trust can be placed on the equipment. The idea is that the user community will depend on this information. Disruptions to the data flow, in this case the identification, are very costly. The reconciliation is fraught with manual efforts, and usually have significant time delays.

Reliability is expressed in terms of a per cent. Most of the standards have used a reliability of one error in ten thousand or 99.99%. Some customers have said that their manual processes have been able to provide the correct identification 85% of the time. While the comparison is not totally accurate, the 85% is 1,500 errors in ten thousand. This is a dramatic difference. Some wonder why the comparison is not fair, but the component of the radio frequency equipment is 99.99% versus the total system being used in the second case which offers 85% reliability.

RELIABILITY EXAMPLES

PERCENT	ERRORS PER 10,000
99.99	1
99.00	100
95.00	500
85.00	1,500

CHART: RELIABILITY EXAMPLES

Note that the chart on reliability examples shows the one in ten thousand for 99.99%. But look at the 99.00% level. Here there are 100 errors per ten thousand. If you processed 2,000 things a day that would mean that on average you would have 20 errors if the system provided 99.00% reliability. Compare this to the one per week in the previous example. Of course, there would be 300 errors per day if the operation was running at 85%. Think of the effort required to correct these errors and the potential for loss of goodwill.

ACCURACY

Accuracy refers to a process being without error or mistakes. This means that when the system reads the identification, it will be the identification of the transponder being read. The system is to produce the result rather than appear to produce the result. The specifications for this area are much more demanding than the specification for reliability. Again, this specification is expressed in terms of a percentage. The specifications are generally 99.9999% for accuracy. This means that when the system produces an identification, one time in a million times the identification presented will NOT be the one container in the transponder.

TAG LIFE

Tag life refers to how long the tag will last when the tag is used under normal conditions. Normal conditions include the environmental conditions that are part of the specification. An abnormal circumstance might be a situation where something is trying to destroy the equipment beyond what was considered normal in the specification. A tag could be mounted on a car and travel with it in the course of driving and this might be considered normal. The same tag taken off the car, hit repeatedly with a sledge hammer, and subjected to the heat inside a blast furnace, might not be considered normal.

Tag life has to do with the packaging and environmental specifications. It also has to do with the life of any batteries that are used to power the transponder whether it's in active or trickle mode or not.

SPEED

This refers to the speed of the tag and reader or interrogator in relation to each other. Generally speaking one or the other is in a fixed location and the other is moving at the specified speed.

There are also speed specifications for the write operations of read/ write systems. These speeds do impact the application capabilities that are possible.

TEMPERATURE

The specifications for temperature involve the operating and storage limits. Because the transponder and read / write equipment are different, and perform different functions, there may be different temperature specifications for each. Temperature specifications for radio frequency identification systems tend to be more dramatic than say a specification for office equipment. Often, it falls into the rugged military specifications for the environment, but a caution here might be that there is an area between an industrial specification and military specification that can serve most needs. As manufacturing processes and materials are made available, it may be affordable to meet the most stringent of specifications. The caution here is that, in general, the more severe the specification, the more costly the product. There may be a practical limit that is more affordable.

FREQUENCY

Frequency refers to the standard transmitting and receiving of radio energy. The energy is transmitted through an antenna and reflected by the transponder. So the frequency used in this identification system is for the primary purpose of having the reader /interrogator communicate with the transponder.

The frequency range used comes from the range set aside by the governing bodies for such purposes. North America uses a band of 902 to 928 MHz, while in Europe 2400 to 2500 MHz is used. The frequency ranges for a given system will change as the governments decide on the standard uses of the frequency spectrum. Lower frequency ranges are being considered in Europe. Certain design points for tags provide the ability for the tag to be read in two different countries at two different frequencies.

High frequency tags are used, and can be battery assisted to meet the frequency requirements. These fall into the spectrum of microwaves. The term, low frequency, is used also, and designates frequencies between AM radio and inductive heating.

There are other related specifications such as band width, harmonic output, and spurious output. The power is also specified as well as frequency stability.

TAG LIFE

The tag or transponder life is specified so that the tag will survive and operate properly under conditions of its expected operating environment. Terms like tamper proof, and sealed, come into play here. The transponder must meet appropriate test standards for long term physical, radio frequency, thermal, and ultra-violet exposures. The life is usually specified in terms of years. A 10 year tag life means, on the average, all tags in the population will survive past the ten year limit. So, this is a minimum life expectancy.

TAG POSITION

For those organizations that are concerned about fleets of equipment, or standard containers and pallets, care is taken to insure that the position and orientation of the tag is in the recommended location. This insures the probability of a read is increased by all participating parties of the equipment, and that all equipment that is used together can be tagged and read reliably. So, users of readers can expect a tag to be in a certain location. This ensures that the antenna can be tuned to read the tag in the area where the tag will most likely appear.

For different equipment to be tagged, care is taken so the reading of one tag does not interfere with the reading of others. Depending on the tag type and reader protocols used, a polarity scheme is used, or other communication protocol is taken into account. One example is to place a tag in the horizontal plane on equipment A and in the vertical plane on equipment B.

DATA CONTENT

The standards also organize the representation of the data fields to assist in maximizing the information flow, and to insure consistency of use between organizations. Things like type codes, field sizes and user fields are specified.

DISTANCE

Specifications are also given showing the minimum and maximum read and write distances.

SPECIFIC STANDARDS

The detail of the specific standards can, of course, be found in the documentation provided by the standards organization. The organization should be contacted to get the latest version for your use. Along with the standards are recommended practices that will provide additional information for the user community.

Before reviewing some of the specific standards written for automatic equipment identification using radio frequency equipment, it is worth understanding the concept of referencing other standards within a given standard. For example, there is a set of standards for freight containers that encompass the coding, identification and marking for freight containers. A standard for freight containers using RF/ID might refer to the ISO 6346 - Freight Containers-Coding, Identification and Marking.

MILITARY STANDARD 810-D

The military has specified many standards that can be used commercially. Military Standard 810-D, Environmental Test Methods and Engineering Guidelines is one pertinent to RF/ID. The guidelines are specific sets of tasks and conditions that, if met, will provide a product that will survive in the likely environment. So, the military specification and method is described for the conditions that need to be met. Here are some examples.

Suppose that a tag is to survive and operate through the contaminants, shock and vibration experienced in rail service and highway and maritime service. Then the tag shall meet or exceed the current version of the following environmental standards.

LOW TEMPERATURE	Mil STD 810D Method 502.2; minimum temperature of -50 degrees C.
HIGH TEMPERATURE	Mil Std 810D Method 501.2 Procedure II; cycled between +70 and +38 Degrees C.
MECHANICAL SHOCK	Mil Std 810D Method 615.3 Procedure I; 30g for 11 milliseconds, half sine pulse.
RANDOM VIBRATION	Mil Std 810D Method 520.0, Procedure II; With TEMPERATURE two hour duration/axis up to 3g at -50 degrees C, ambient, and +70 degrees C ambient.
HUMIDITY	Mil Std 810D, Method 507.2; 95% non-condensing.
RAIN	Mil Std 810D, Method 506.1, Procedure II
SALT FOG	Mil Std 810D Method 509.2, Procedure I
DROP SHOCK	Mil Std 810D Method 516.3, Procedure II; Height 3.3 meters, impact surface 5 cm plywood backed by concrete.
LEAKAGE	Mil Std 810D...
ICING/FREEZING RAIN	Mil Std 810D...
SAND and DUST	Mil Std 810D...

When a tag goes through this kind of testing, and passes, it has been cooled, heated, dropped, shocked, submerged in water, subjected to corrosive salt, fog, blasted with sand and dust, subjected to ice and cycled through many of these actions. These actions all simulate the environment where the tag is to operate. Our confidence factor is higher after the successful completion of these tests. Of course, multiple tags are put through the tests just to be sure.

For the uninitiated, the processes might seem obvious, trivial or extremely complex. Having witnessed several tests on transponders and hand held computers, I have found the testing process rather rigorous. The tests require specific test bed conditions, so the tests are not performed just anywhere. Specific testing labs are certified as capable of performing the tests. Special care is taken to adjust the machines necessary to drop the test item.

As an example, a simple corner drop might be the test process. Care is taken to insure that the object will fall from the correct height. Special devices are used to hold the device in place to insure that the object will fall on the corner to be tested. If all corners are to be tested then there are likely to be four positions for the drop. A specific drop test may test a different item for each corner. Other tests may require one item with four drops. The surface is tested as well. I was surprised to find out that a drop on a cement floor required a specific type of cement that will have the proper hardness characteristic. In the test described above, the plywood was 5 cm thick. Also the plywood has other specifications before it is used. The cement, that the plywood lays on, also has to meet specific requirements.

One of the elements sometimes glossed over is the type of dust and sand used in the contamination tests. It is not the kind you and I might be very familiar with, because of our household duties or trips to the beach. No, this element has particular specifications, like particle size and chemical makeup. The dust used in most quality tests is manufactured, and then sold sometimes for hundreds of dollars per pound.

All of this emphasizes the critical nature of the specified tests. There is a point to made by the test, and a protection implied, by a passing or failing grade.

AAR STANDARD FOR AEI

The Association of American Railroads Standard specifies requirements for the automatic electronic identification of equipment used in rail transportation, such as rail cars, locomotives, intermodal vehicles and end-of-train devices. The standard describes a reflected energy system. Sensing equipment decodes radio waves reflected by a tag mounted on the equipment (described above) used in the transportation industry. The radio waves indicate the identification code of the equipment as well as its related permanent information.

The identification system and data output described is used to identify equipment by the initial and number, or the individual alpha-numeric marking and other predefined information.

AAR AND OTHER STANDARDS

The system and the data output described in the standard are compatible with the ANSI Standard, MH5.1.9-1990, and the ISO Standard, 10374, for the automatic identification of containers. The standard also complies with the ATA Standard for automatic identification of trailers and chassis. The ATA Standard also covers other highway equipment such as, tractors, straight trucks, and converter dollies.

AAR GENERAL REQUIREMENTS

For automatic equipment identification purposes, each unit of equipment is fitted with a transponder, or small tag, containing the alphanumeric code of the equipment and the related information. This code is to be read by an interrogator which operates on ultra high frequency radio waves. This interrogation reads or decodes the altered radio waves reflected by the transponder on the equipment. The altered radio waves indicate the alphanumeric identification code of the equipment, as well as other predefined information.

The interrogator adds its own identification number, the date and time, and transmits all of this data over the user's communications link used for sending messages. The user provides the communications line.

The system is expected to accurately read freight trains moving at up to 80 miles per hour. This includes configurations such as double stack containers, including 20 foot units, containers on chassis, on flatcars, and end-of-train devices. This requirement applies in areas of one, two, or more parallel tracks, at ordinary centers, with trains standing or operating on any, or all, of these tracks, in the same or opposite direction.

AAR TAG REQUIREMENTS

Tags must be approved in accordance with specified standards and approval procedures. The tag must be tamper proof and sealed so that it will survive and operate properly under the conditions of its expected operating environment. The tag life must not be less than 15 years, and no maintenance should be required. The tag must meet appropriate test standards for long-term physical, radio frequency, thermal, and ultra-violet exposure.

There may be two versions of the tag; battery-powered and non-battery-powered. The advantages of the battery powered tag includes a greater range and a reduced power requirement from the reader system. The advantage of a non-battery-powered tag is a longer life.

TAG MOUNTING LOCATION

The tag mounting surface must be metal, vertical and smooth within the area of the tag. No area of the tag's rear surface may be more than 1/4 inch from the metal mounting surface. In case the desired mounting area will not meet this requirement, a mounting bracket must be provided to satisfy this requirement. If the mounting surface is not metal, as in the case of fiberglass, then a metal back plate or mounting bracket must be supplied to satisfy this requirement. A 1/8 inch or thicker metal back plate extending from each side of the tag should be used. Refer to Figures: Tag Mounting Clearance Zone and Mounting Location Examples for additional clarification.

AAR INTERROGATOR REQUIREMENTS

No minimum or maximum interrogator power is specified. However, the minimum antenna effective isotropic radiated power and interrogator receiver sensitivity must be adequate to properly interrogate tags, capable of responding as specified in this document at all distances between the minimum and maximum distances specified by the user. The maximum effective isotropic radiated power and transmitter power output of the interrogator must be within the limits prescribed by the telecommunications authority of the county in which the interrogator is operated.

Interrogator units must be capable of interrogating multiple tags within their reading field, and discriminating between the tags without misreading. Interrogators employing tag response levels as a method of discriminating between multiple tags may accomplish this by distance differential and/or position relative to the antenna pattern. Error detection ensures reading accuracy.

The system detects the presence and direction of the movement of each unit of rolling stock and properly identifies rolling stock without tags or with incorrect tags. The system must detect the presence of equipment on the track. A function of the system is to activate the system when a presence is detected. A clean list application containing no duplications of car initials and numbers, as well as, identified "non-tagged cars" must be activated when the idle mode or nothing present is indicated.

AAR RAILCARS AND LOCOMOTIVES

Each railcar must have two tags, one on the B end, left (BL) and the other on the A end, right (AR) portion of the car. The locomotive must have two tags, one located at the F end, right (FR) portion and the other at the R end, left (RL) portion. Optionally articulated cars can carry two tags on each platform.

FIGURE: TAG MOUNTING CLEARANCE ZONE

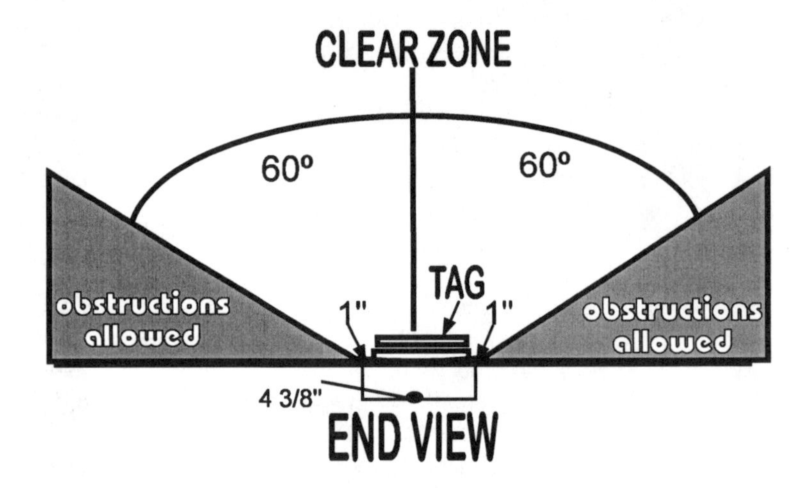

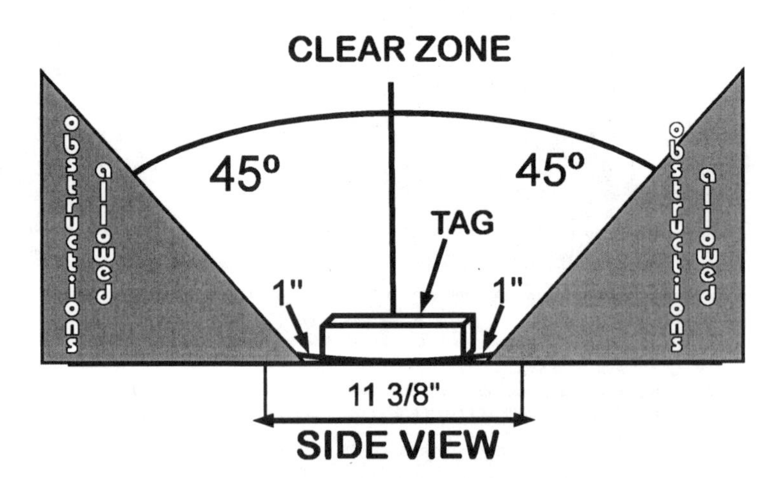

Equipment owners, who cannot reasonably place tags within the limits of the specified location window described in the standard, must send in a written request for a variance. The request must include detailed information, including equipment initial and number, car builder and equipment model. Diagrams, drawings and photographs should accompany the request.

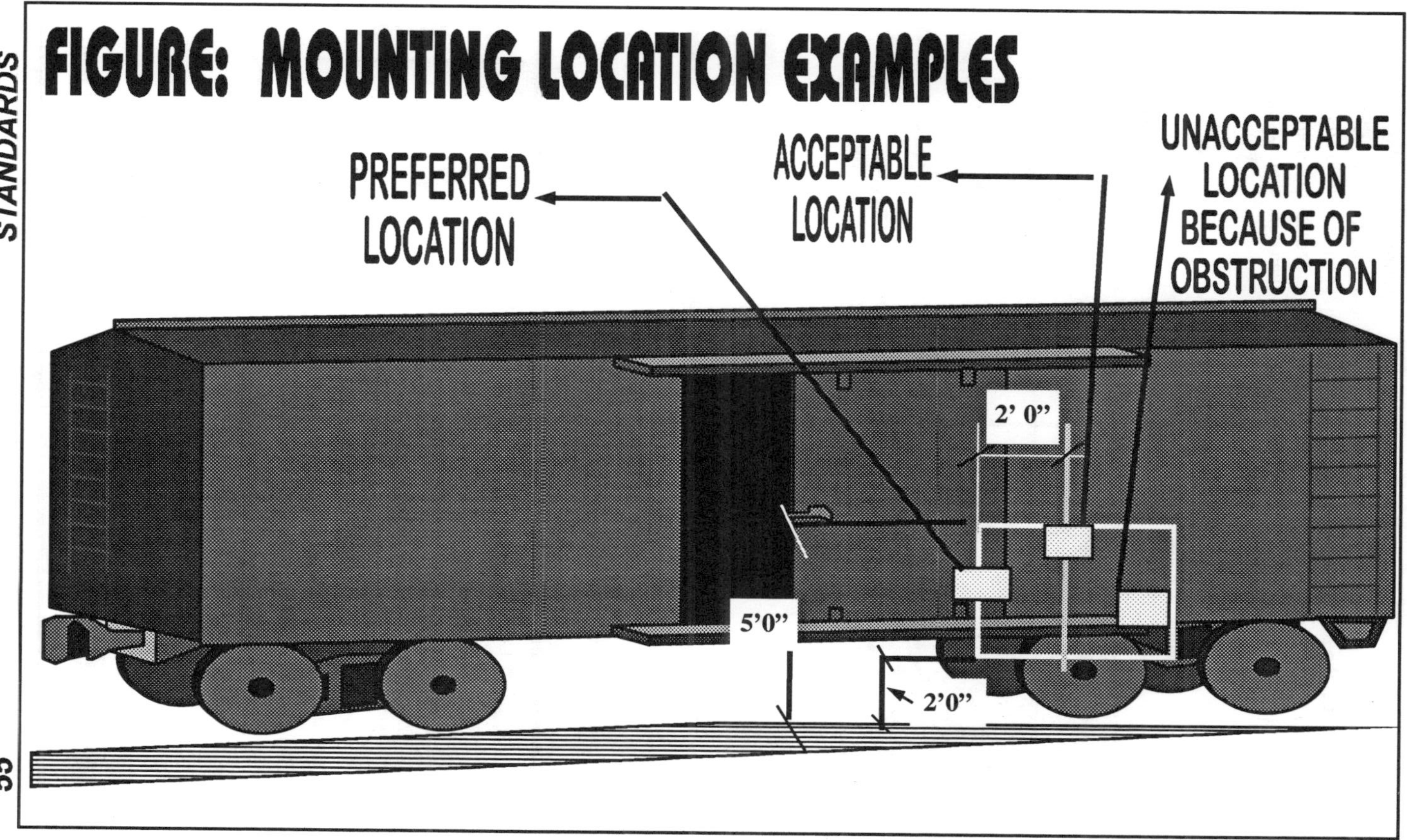
FIGURE: MOUNTING LOCATION EXAMPLES
PREFERRED LOCATION
ACCEPTABLE LOCATION
UNACCEPTABLE LOCATION BECAUSE OF OBSTRUCTION
2' 0"
5'0"
2'0"

END-OF-TRAIN DEVICES

Each EOT can have one tag, and the option of two tags. The owner determines the side.

AAR CONTAINERS

For containers 40 feet in length or less, the equipment tag should be located on the forward right sidewall of the container approximately one foot to the rear of the front corner post. For containers that exceed 40 feet in length, the tag must be adjacent to the rearward side of the post at the 40-foot corner lock position.

AAR CHASSIS

The tag should be located on the right hand front corner of the forward bolster. The tag must be oriented in a forward direction and should respond to a horizontally polarized interrogating signal.

AAR TRAILERS

When possible, the tag should be located on the forward right sidewall of the trailer approximately one foot to the rear of the front of the trailer, centered one foot below the roof line. The polarization shall be vertical.

AAR DATA FIELDS

To understand the data fields the following table is offered.

TABLE: AAR DATA FIELDS RAILCAR AND LOCOMOTIVE

DESCRIPTION	CAR	LOCOMOTIVE	
Equipment group code	5	5	
Tag Type	2	2	
Equipment Initial (Mark)	19	19	
Car/Locomotive Number	20	20	
Side Indicator Code	1	1	
Length	12	9	
Number of Axles	5	5	
First Check Sum	2	2	
Reserved Frame Marker	2	2	
Bearing Type Code	4	3	
Platform Identifier Code	4	30	spare owner use e.g. model no.
Spare #1 Owner Use	5		
Spare #2	10		
Spare #3	7		
Reserved AAR USE	9	8	
Security	12	12	
Data Format Code	6	6	
Second Check Sum	2	2	
Frame Marker	2	2	
Total Bits	128	128	

This fixed length record concept truly is an identification system. The fields vary by type but nonetheless indicate the permanent information stored within the tag. There is a set of user entries available, but the information must be fixed into the storage of the transponder. As you can see by the two types shown, there is a lot more information than the car initial and number.

CHAPTER VI

ISO FREIGHT CONTAINERS - AUTOMATIC IDENTIFICATION

ISO

The International Organization for Standardization, (ISO), is a worldwide federation of national standards bodies (ISO member bodies). The work of preparing International Standards is normally carried out through ISO technical committees. Each member body interested in a subject, for which a technical committee has been established, has the right to be represented on that committee.

ISO FREIGHT CONTAINER STANDARD

This International Standard specifies a system for the automatic identification of a freight container and permanent related information to send to third parties in a standard format. It is intended that the Automatic Equipment Identification (AEI) system will facilitate documentation, resource control, and communications (including electronic data processing systems). The visual container identification markings specified by ISO 6346 are not affected. Future additions of this International Standard will specify modulation, encoding and an open protocol.

Annex B, which is an information annex only, describes the technical specifications of a system that complies with the requirements of this International Standard. Amtech Corporation of Dallas, Texas holds patents for parts of Annex B.

ISO SCOPE

The scope of the standard covers many aspects for electronic and automatic identification of containers. It specifies all necessary user requirements in order to permit international use without modification or adjustment. The International Standard applies to all freight containers as defined in ISO 668.

The use of AEI systems and the equipping of containers for automatic identification is not mandatory. The purpose of this International Standard is to optimize the efficiency of equipment control systems. Following are the key points of the standard.

1. Container automatic electronic identification.
2. Tag data coding system for ID and pertinent other data.
3. Data coding transfer from tag to processing system.
4. Description of data in the tag.
5. Performance criteria.
6. Physical location requirements.
7. Security features

ISO OPERATIONAL REQUIREMENTS

The AEI system shall consist of two basic components.

1. An electronic device called a tag that is installed on the freight container.
2. Electronic sensing equipment located apart from the freight container.

ISO TAG

The tag MUST be capable of maintaining the integrity of the freight container identification and permanent related information. The data will be programmed into the tag in the field, but not reprogrammed while the tag is affixed to the container. The tag must be physically and electronically secure and tamper-proof. The dimensions are not to exceed 30 cm x 6 cm x 2 cm. The life expectancy should be 10 years with normal operation and must not require periodic maintenance. It also must be capable of international operation, without the necessity of licensing tags individually.

The tag provides, as a minimum, the basic information about the container so that reference to other tables and databases is unnecessary. The following basic tag information is mandatory and permanent:

A. Tag type
B. Equipment identifier
C. Owner code; ISO 6346
D. Serial Number, ISO 6346
E. Check digit, ISO 6346
F. Length
G. Height
H. Width
I. Container Type code, ISO 6346
J. Maximum gross mass
K. Tare mass

The information contained in the tag is in one, or more, of the following categories:

A. Mandatory, permanent (non-changeable) information
B. Optional, permanent (non-changeable) information
C. Optional, non-permanent (changeable) information

Optional information contained in a tag should not adversely affect the operation of systems requiring only the mandatory information contained in the tag.

ISO SENSING EQUIPMENT

The sensing equipment must be capable of reading the information contained in the tag when the tag is properly presented. The sensing equipment must be able to decode the information for transmission to automatic data processing systems.

The sensing equipment and the connected real time electronic data processing (EDP) system must be capable of adding to the tag data the following information:

A. Sensing Equipment Unit identification

B. Date and Time

C. Freight container movement status

The sensing equipment should be a technology adaptable enough to accommodate fixed or mobile installations, or portable applications.

ISO "PROPER PRESENTATION" OF THE TAG

The orientation requirements for "proper presentation" of the tag to the sensing equipment are illustrated in the figurc: Tag Presentation Requirements on pagc 61. Four tags are labeled A, B, C, and D.

A tag is considered as "properly presented" in terms of its orientation, even if it has an angular displacement such as illustrated with Tag A. The tag might be presented, rotated about an axis, perpendicular to the face of the tag by an amount not to exceed 20 degrees to either side of the vertical.

The tag is required to perform satisfactorily if the interrogating signal reaches it from any direction within a specified cone formed from a line from the center of the sensing equipment and the center of the tag.

Although the tag is required to perform satisfactorily if the interrogating signal reaches it from any direction within a specified cone representation, the "window" within which a tag must be presented to a particular sensing device will vary with the design of the sensing equipment.

The AEI system must be capable of reading a tag on containers in accordance with all the combined requirements given in the following table.

PASSING SPEED	RANGE	DISCRIMINATION	ANGLE
130	1-13 M	10M	20
80	1-13 M	5M	30
30	1-10M	1.2M	70
0	0.1-2M	1.5M	90

TABLE: READ REQUIREMENTS COMBINED

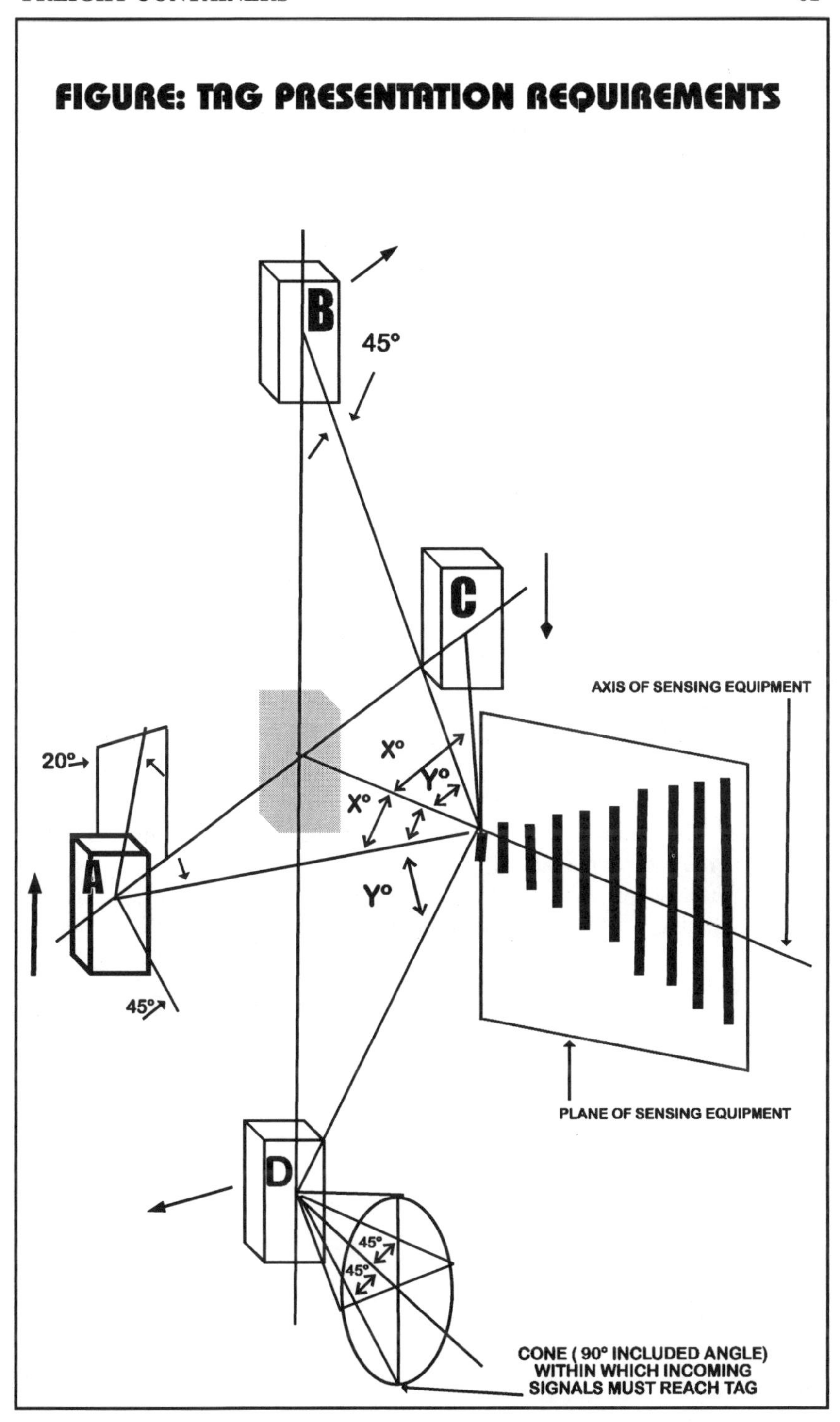
FIGURE: TAG PRESENTATION REQUIREMENTS
B
45°
C
AXIS OF SENSING EQUIPMENT
20°
X°
Y°
X°
A
Y°
45°
PLANE OF SENSING EQUIPMENT
D
45°
45°
CONE (90° INCLUDED ANGLE)
WITHIN WHICH INCOMING
SIGNALS MUST REACH TAG

ISO ANNEX A

Annex A describes the tests required of the tag and sensing equipment. The tag is to be tested based on Mil. Std 810D as described in the documentation.

ISO ANNEX B

This annex describes a reflected energy system in which sensing equipment decodes radio waves reflected by an identification tag mounted on a freight container. The reflected radio waves indicate the alphanumeric identification code of the container as well as its related permanent information. The sensing equipment must transmit a continuous wave carrier signal at a user frequency within either the 850 MHz to 950 MHz band, or the 2400 MHz to the 2500 MHz band, as may be allowed under national and local regulations. Annex B, which is an information annex only, describes, in part, material covered by patents held by Amtech Corporation of Dallas, Texas.

AMERICAN NATIONAL STANDARDS INSTITUTE

ANSI

An American National Standard requires approval. A verification process is followed before the standard is approved by The American National Standards Institute, ANSI. An ANSI Board of Review requires consensus of all views and objections and that a concerted effort be made toward their resolution. The standards may be revised or withdrawn at any time. Contact the ANSI group to get the latest information.

The American Institute of Merchant Shipping sponsored the standard for Freight Containers - Automatic Identification. The approval came in October of 1990.

The ANSI standard, MH5.1.9.1990, is similar to that of the ISO 10374 for Freight Containers Automatic Identification.

CHAPTER VII

ATA STANDARD FOR AEI

The American Trucking Association Standard specifies requirements for automatic electronic identification of equipment used in road transportation. The equipment includes tractors, trailers, dollies, intermodal containers, and intermodal chassis. Described is a reflective energy system in which sensing equipment decodes radio waves reflected by an identification tag or transponder mounted on the equipment used in the transportation industry. The reflected radio waves indicate the identification code of the equipment, as well as its related permanent information. This particular standard is not a requirement for acceptance in road interchange service.

For automatic identification purposes, each unit of equipment is fitted with a small electronic device, a tag, containing the alpha-numeric marking or identification code of the equipment and related information. When the tag is in the presence of sensing equipment, or the reader, operating on ultra high frequency radio waves, the tag reflects altered radio waves. The reader decodes the altered radio waves to determine the alpha-numeric identification of the equipment as well as other predefined information, which is permanently encoded and resident in the tag.

The reader can optionally add its own identification number. The date and time can also be added. All of the information is then sent to the user's computer system. The system is expected to accurately read tagged tractors, moving at up to 65 miles per hour. The system requires only one tag per equipment unit.

ATA TAG REQUIREMENTS

The tag must be tamper-proof and sealed in such a way that it will survive and operate properly under the conditions of its expected operating environment. The tag life must not be less than 5 years and no maintenance should be required. The tag must meet appropriate test standards for long term physical, radio frequency, thermal and ultra violet exposure. The tag shall not be damaged by the normal operation in and near roadway, railroad, marine or other distribution facilities.

The tag must operate properly, within the temperature range of -45 degrees C to +85 degrees C. The tag must maintain the integrity of stored data at temperatures of -60 degrees C to +85 degrees C.

The tag should survive and operate through the shock, vibration, and chemical contaminants experienced in road service. This is per Military Standard 810-D.

ATA READER REQUIREMENTS

No minimum or maximum reader power is specified. However, the minimum antenna effective radiated power and reader sensitivity must be adequate to reliably record the passage of properly presented tags at distances compatible with typical recording locations such as: entry/exit gates, multi-lane highway checkpoints, weigh stations, terminals and yards, fuel islands, and maintenance facilities. To accommodate these facilities, the reader must read a properly presented tag at distances of at least 42 feet (13 meters) from the reader at speeds up to 65 miles per hour.

ATA TAG FREQUENCY

TAG TYPE	FREQUENCY MHz
General	902 - 928
Container	850 -950 2400 - 2500

ATA SYSTEM OPERATION

The radio communication system described consists of a reader system (reader, RF Module, and antenna) and tags. Tags are placed on objects to be identified, and reader systems are installed at points to record the passing of tagged objects. The system is designed for localized application, where the tag passes by the reader system. It should be noted that either the tag or the reader may be moving.

ATA TAG PROGRAMMER

This device is used to place the coded information on the tag. When security is being used, a unique security value is given to the programmer for automatic inclusion in the tag coding scheme. The programmer automatically calculates check sum fields.

ATA DATA CONTENT AND FORMAT

The tag is composed of a minimum of 128 bits of non-volatile memory which can be divided into two sections. The first section is composed of data bits which are allocated for procedural needs and the second section is composed of data bits which are available for general use. Procedural needs include error checking, detecting the beginning and end of the tag's 128 bit data stream, indicating the type of data format utilized in the tag, and providing security from unauthorized duplication of tags. Twenty six bits are used for procedural needs and 102 bits are available for general use.

It should be noted that the 12 security bits are reserved for security purposes. However, if security is not desired, these bits can be designated for limited general use. The following table shows the bits reserved for procedural needs.

TABLE: PROCEDURAL NEEDS

FIELD DESIGNATION	BIT POSITION
FIRST CHECK SUM	60, 61
RESERVER FRAME MARKER	62, 63
SECURITY	106 - 117
FORMAT CODE	118 - 123
SECOND CHECK SUM	124 - 125
FRAME MARKER	126, 127
POSSIBLE BIT POSITIONS	**0 - 127**

The fields are arranged in a hierarchical fashion in order to expedite translation and processing by the data processor. It is intended that the data processor will first look at the Data Format Code to determine if the tag should be decoded or ignored. For example, in some cases the data processor will wish to ignore all tags except those specified as highway (ATA Standard) or marine intermodal (ISO Standard).

Once the data format code has been processed, then the data processor will look to the tag type to determine the configuration, capabilities, and memory capacity of the tag.

Next, the data processor will examine the equipment group code to determine if the tagged equipment is relevant. For example, the processor may ignore or process non-revenue equipment differently than it would trailers or dollies.

The Table: ATA DATA FIELD DESCRIPTION SUMMARIES is provided to view the fields and descriptions to gain an appreciation for the information contained in the standard. Detailed descriptions can be found in the standards documentation from the ATA. Notice that the data formats have similarities, but have been designed so that the pertinent information about the equipment is shown for the different equipment types. There are also some reserved and spare areas for use as the application matures and new features are required.

ATA LOCATION AND MOUNTING OF TAGS

The tag should be capable of permanent mounting and have normal dimensions which do not exceed 30 x 6.0 x 2.0 cm. The tag attaches to a metal surface or metal plate that is then attached to the equipment. The metal surface should be flat and have a surface area which exceeds the surface dimensions of the tag by at least 50%.

TABLE: ATA DATA FIELD DESCRIPTION SUMMARIES

DESCRIPTION	TRTR	DLLY	TRLR	CHAS	ICNR
Tag Type	5	5	5	5	5
SCAC Code	2	2	2	2	2
Identification number	19	19	19	19	19
Chassis Type Code				4	4
Tare Weight				6	11 check digit length
First Check Sum	2	2	2	2	2
Reserved Frame Marker	2	2	2	2	2
Number of Axles	3	3			
Length			11		
Height				7	9
Width Code			2	2	7
Trailer Type Code			4		7
Max Gross Weight					9
Forward Extension			8	6	
Tare Weight	8	6	7		7
Wheelbase	6				
Fifth Wheel Offset	4				
Tare Weight Steering Axle	5				
Drive Axle Spread	5				
Reserved	3				
Dolly Type Code		7			
Spare		7			
Reserved		12			
Kingpin Setting				6	
Axle Spacing				5	
Running Gear Location				5	
Number of Lengths				3	
Minimum Length				10	
Spare				2	2
Security	12	12	12	12	12
Data Format Code	6	6	6	6	6
Second Check Sum	2	2	2	2	2
Frame Marker	2	2	2	2	2
TOTAL	**128**	**128**	**128**	**128**	**128**

KEY: TRTR = Tractor DLLY = Dolly TRLR = Trailer CHAS = Chassis ICNR = Intermodel Container

ATA TRACTOR

The tag should be mounted on the front surface of the tractor on the right hand side in the vicinity of the front bumper. It may be mounted in the placement window which extends horizontally from 30cm (1 foot) to 90cm (3 ft.) to the right of the bumper's center line and extends vertically from 60cm (2 ft.) to 105cm (3.5 ft.) above the ground. The tag must face forward and be installed horizontally to respond to a horizontally polarized signal from the reader system.

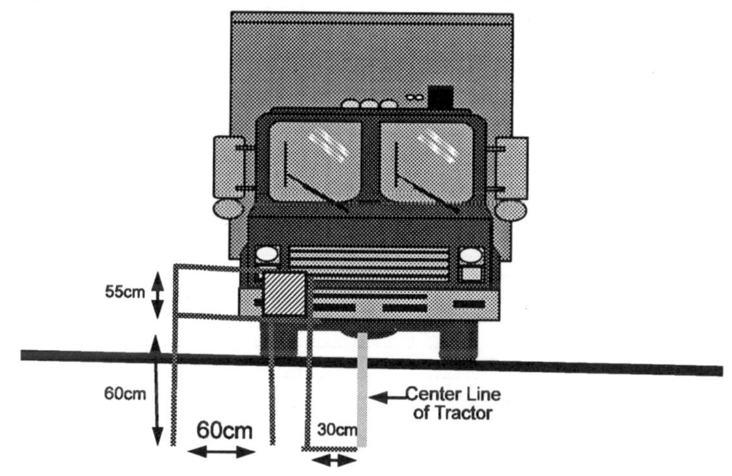

FIGURE: TRACTOR TAG PLACEMENT

ATA DOLLY

The tag must be positioned on the right extension arm, facing outward and located midway between the axle and the pintle hook with a tolerance of +20cm toward the hook. A horizontally polarized position is required.

FIGURE: ATA DOLLY TAG PLACEMENT

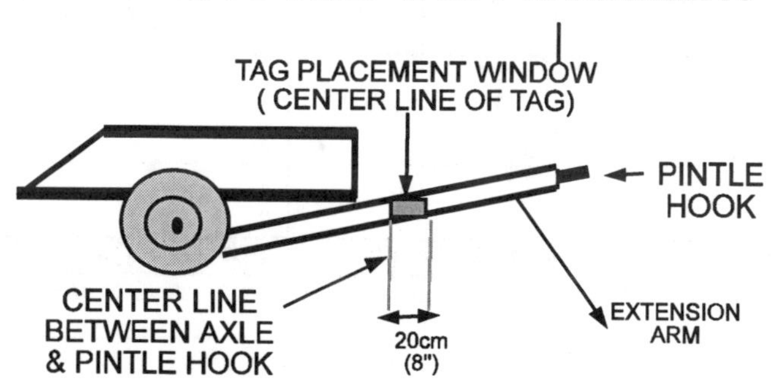

The tag must be located on the forward right sidewall of the trailer approximately 30 cm. (one ft.) below the roof line. Mount the tag so that it will respond to a vertically polarized signal from the reader system.

For flatbed trailers and other situations where the above mounting location is not available, an alternative trailer tagging position is provided. This tag shall be located on the front surface of the trailer on the right side.

The tag may be located in a placement window which extends horizontally from the trailer's right side to a point 60cm (two ft.) toward the center of the trailer and extends vertically from the bottom surface to the trailer to a point 30cm. (one ft.) above the bottom surface.

The tag must face forward and be positioned horizontally to respond to a horizontal polarized signal from a reader system.

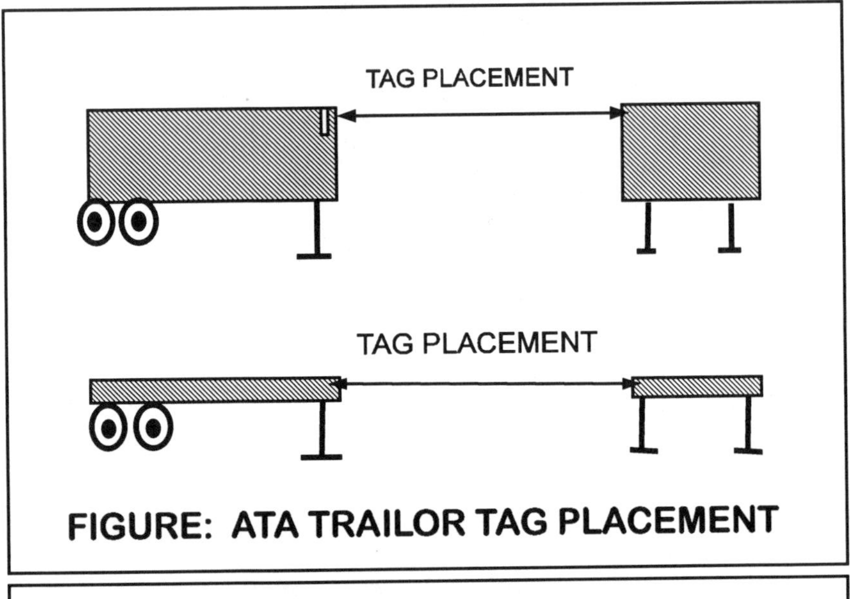

FIGURE: ATA TRAILOR TAG PLACEMENT

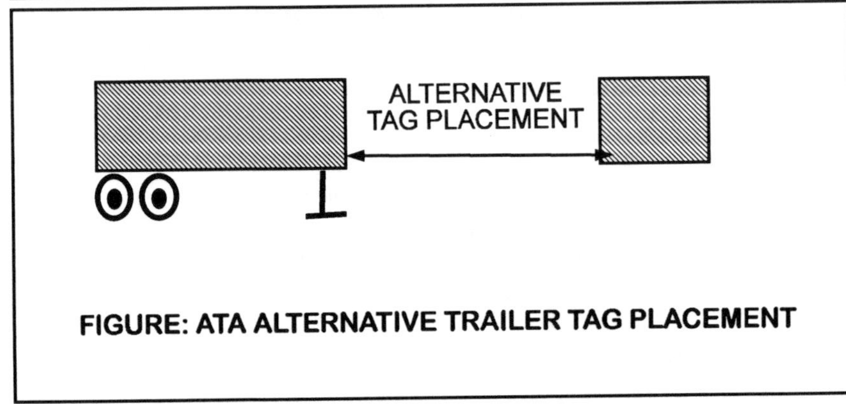

FIGURE: ATA ALTERNATIVE TRAILER TAG PLACEMENT

ATA CHASSIS

The tag should be mounted in the right hand front corner of the forward bolster. The tag must face forward. The tag is mounted and read in the horizontal orientation.

ATA CONTAINER

For containers 40 feet in length or less, the equipment tag should be located on the forward right sidewall of the container, approximately one foot to the rear of the front corner post within the first corrugation (if possible), centered one foot below the roof line of the vehicle. Containers that exceed 40 feet in length, the tag must be adjacent to the rearward side post, at the forward 40 foot corner position.

CHAPTER VIII

HUGHES PROTOCOL

The following chapter provides a description of the Hughes protocol and it's potential uses for the technology.

TWO-WAY VEHICLE TO ROADSIDE COMMUNICATIONS EQUIPMENT

Surface transportation is a significant opportunity for the application of technology. Highways are congested and space to build more lanes is not available. Traffic accidents are of prime concern for the safety of drivers and passengers. Repair actions on the vehicles drain away billions of dollars each year and estimates of $100 billion each year in lost productivity are thought to be low. More effective use of the public and private roadways is critical to safety, health and productivity issues. The Intelligent Vehicle Highway Society of America, IVHS, has specified applications and projects that can implement the improvements needed. Technology is needed to provide information to the authorities, improving utilization and increasing traffic flow. One such technology involves vehicle to roadside communication equipment.

Vehicle to roadside communication can be thought of in two primary application areas:

1.Collection
2.Status

The toll collection applications have been in use for several years in the read-only mode, using a radio frequency identification system. This has proven to be very productive in terms of lowering queue times for the vehicles at toll gates. Status information has a wide range of possibilities, including traffic flow indications, and regulatory information, such as weigh scale applications. More applications like traffic conditions and routing are possible.

PROPOSED STANDARD

History has shown, with regard to radio frequency identification systems, that there has been vendor positioning of proprietary solutions. The Hughes Aircraft Corporation has developed a working prototype that offers new function to meet the IVHS objectives, while offering the protocol in the public domain. One feature is the "Slotted Aloha" protocol, and there are other functionally rich innovations. This standard opens up for the competitive market the opportunity for many players and a chance for wide acceptance. A standard is proposed and while the process may change some of the specific features. The following is intended to provide a general overview of how the system works.

SCOPE

The standard defines the data link protocol and the radio system for short range, half-duplex, active,and two-way vehicle to roadside communications equipment. Short range, in this definition, is 75 meters or so. Half duplex means that the two ends of the system, the transponder and the reader, send in one direction at a time. Active means that the transponder, when triggered by a radar control message, transmits the requested information. This is different from modulated backscatter systems that reflect and modulate the carrier transmitted by the reader. This latter technique is sometimes called passive technology. Accurate and valid message delivery is critical to the success of the system. Both wide area and lane-based communications are featured.

WIDE-AREA COMMUNICATIONS

The wide-area protocol permits transactions with several vehicles travelling on a multiple lane roadway without restricting the vehicle to any fixed lane, trajectory or speed. The application may be characterized by the capability to perform general two-way digital communications with multiple vehicles simultaneously in an open road operating environment, with minimal implementation restrictions.

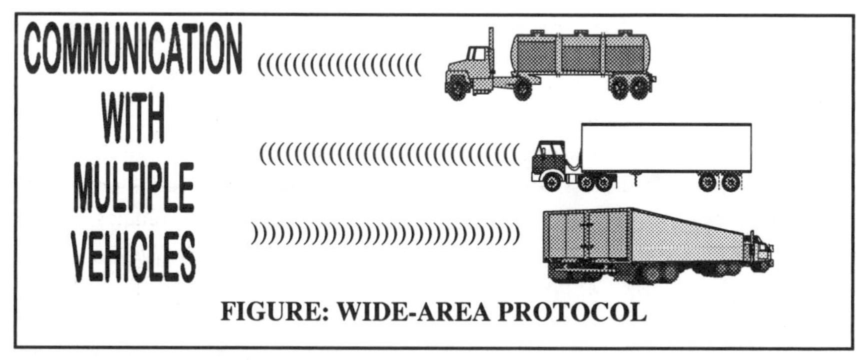

FIGURE: WIDE-AREA PROTOCOL

A clear line of sight is required from the transponder to the reader's antenna for all transponders in the field of view. The reader controls all the transactions with multiple transponders travelling unconstrained on the roadway. A Slotted Aloha protocol is implemented. Slotted Aloha is a Time Division Multiple Access, (TDMA), protocol based on a cyclic structure, known as a frame. The frame consists of a reader message, a transaction cycle, and an activation cycle. The protocol permits multiple transponders to simultaneously request permission to perform a transaction. The reader then commands transponders to communicate in specific, dedicated time slots within the frame. At the conclusion of each transaction, a positive confirmation is performed. If the transaction fails for any reason, a mechanism to repeat the transaction is initiated.

Time division systems are often subject to non-productive collisions as the population of messages increase. To assist in minimizing interference, the transponder broadcasts in one of 16 random activation slots. If two transponders happen to be broadcasting in the same slot and thus collide causing interference, in the next cycle they will broadcast at two new random activation slots. In addition, the reader can control the time to respond as a percentage of time. As an example, the reader could change the transponder response time from say 25% of the time, to a response time of 12% of the time. This, in effect, reduces a particular transponder's chance of communicating, but provides an opportunity for all transponders in the field of view, an increased opportunity to enact a transaction. Once the last frame transaction is received, there is a sleep timeout sent to the transponder.

LANE-BASED COMMUNICATIONS

The lane-based protocol permits a transaction with a single vehicle travelling on a restricted trajectory. The application is an exchange of a fixed length message in a short duration with a vehicle when the vehicle passes through a specific location on the roadway.

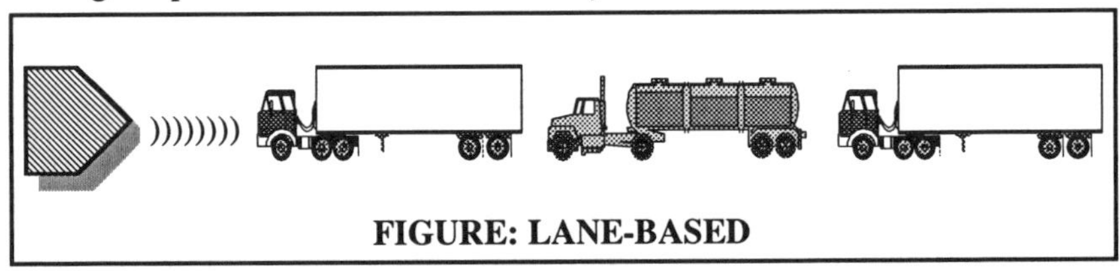

FIGURE: LANE-BASED

In lane based operations, the specific application is a factor in the overall design. The maximum speed of the vehicle, number of lanes being controlled and the expected time the transponder is in the communications zone of the reader, will influence design specifications. Lane-based protocol has four activation slots to handle multiple vehicles in the lane. With the four activation slots there is an increased chance of reading the transponders even when there are tags being shadowed by other tags. A clear line of sight is needed between the transponder and the reader's antenna. The antenna may be in the ground or overhead.

READER

A reader can be thought of as a fixed position controller. The transmit and receive antenna(s), modulation and demodulation hardware, and software are part of the system. The reader type is indicated with a binary code of 1000 for wide-area operation, and 0000 for lane-based.

TRANSPONDER

The transponder is an electronic device attached to a vehicle, and containing information that can be communicated to the reader. Each transponder should have a unique 32 bit code or serial number, and a 32 bit private ID for security purposes. A transponder type code indicates a lane-based transponder or a wide-area transponder. Typically, transponders respond to either or both lane-based and wide-area requests. Transponders can also be classified as read- only, read/write, or other designations. The Hughes protocol is not offered in the read- only format, but the system can be used in that way by simply not writing to the tag. Memory in the tag is another matter. The memory scheme for the transponder is made up of three different types.

1. Factory Programmed (Read Only)
2. Agency Programmed (Read Only) (Read/Write by Agency)
3. Read/Write

TABLE: MEMORY CONCEPTS

	MEMORY TYPES	
FACTORY	AGENCY	READ/WRITE
READ ONLY	READ ONLY	READ/WRITE
SERIAL # PRIVATE ID	ID, OTHER	SCRATCH PAD

FACTORY PROGRAMMED

To insure a unique code, there is a designation of factory programmed read- only memory. This would have, among other potential applications, the serial number concept. The concept is to program this memory once, at the factory, and from that point on, this memory would be read-only. Manufacturers could include version codes and other designations for control purposes.

AGENCY PROGRAMMED

This area of the memory accommodates the user with a permanent location for fixed data. Here the user's identification code is programmed into the transponder with a special device. This memory is read- only, so the operation of programming this area is thought to be a one time deal. In addition to the ID, other permanent information like tare weight, model numbers, and account codes can be placed here.

READ/WRITE

This type of memory is thought of as a scratch pad. The reader, with write capabilities, can write to the memory over the air. This memory need not be retained when the transponder is turned off or the power is removed.

HUGHES PROTOCOL: SUMMARY

This protocol has some significant advantages. The ability to operate in multiple modes is important. Factory coded identifiers, with user programmed read -only areas of memory, is a good idea, and the read/write memory offers particular user application advantages. The standard, as made available, promotes the competitive environment with multiple options. This might be the most significant contribution.

SUMMARY

Standards have been put in place and continue to be developed for the RF technology. These standards provide a way for the user community to gain substantial benefits.

CHAPTER IX

APPLICATION OF RF/ID

This chapter describes the general concepts and cautions for RF/ID systems and the general application of these concepts. It is the basic training required to help the application architect design the system for the user's application.

APPLICATION OF RF/ID

The concept of automatic data collection is not new to the user community. Bar code and magnetic stripe systems have been around for decades. The concept of automatic identification continues to offer the benefits of timely and accurate data collection. Data collection is spreading into areas never before possible. While the concepts are not new, the application of radio frequency identification has some new twists that are worthy of additional comment. This chapter will try to hit the fundamental application uses of RF/ID; basic training so that the specific application can be designed by the user. The topics to be addressed are:

- A. Fixed Point Reader Systems
- B. Fixed Station Transponders
- C. Wide Area Applications
- D. Hybrid Applications
- E. Common 'Gotcha's'

The user community must always define the requirements for the desired system. The requirements should be tested against the system design to insure that the desired results can be achieved. The re-engineering of systems and processes can greatly impact benefits. When tools like RF/ID are employed, it is important to view the solution to the problem from the end user's view, not from the perspective of the RF vendor or the technology. This word of caution should not cause alarm, but should emphasize that it is not important that new technology be installed or implemented, but rather that the strategic customer problem is resolved. To that end, it is important to know what the technology can do, so that the resultant re-engineered process performs the desired functions.

It is also significant to know the level of detail required to answer the user's question. If the user's wants and needs are truly articulated, then you can match the technology to fit the need. Consider the following chart, Requirements Vs Technology, showing requirements versus technology. Anytime you attempt this kind of comparison, there is room for error, but there is an important point for the user. If you do not care what passes by, there is no need for technology. However, if you are managing a department store, you may tag items, and you care if any of the items are leaving the entrance. A simple bit on the tag could do here. If you are interested in knowing what items of a significant value are leaving the store, a range of ID's would do. If store, then a specific identification code is required. Also, more control might be required to insure that a specific item is in a specific field of view. Do you want to know if an item is in the store, or on a particular shelf. More specific requests might change the solution requirement, from a technological view. Either solution might be the correct one for the end user.

CHART: REQUIREMENTS VS TECHNOLOGY

REQUIREMENT	TECHNOLOGY
1. ANY PASSAGE	NO REQUIREMENT
2. ALERT IF ANY SIGNAL	ON BIT, SINGLE BIT
3. ALERT IF IN RANGE	ID CODE RANGE
4. ALERT THIS SPECIFIC	ID CODE REQUIRED WITH CONTROL

STRONGEST SIGNAL

Certain RF manufacturers use the concept of locking in on the strongest tag in the field of view. This provides a control assumption that should be used in your application design. What this means is that there is a high probability of reading the closest tag to the antenna. However, prudence dictates that you understand the quality of the tags in the population. For example, if the signal strength from a given set of transponders to the reader have a high signal variance, it is possible to read a tag other than the closest one. It is also possible to read a tag that is being reflected into the field of view over a closer tag. However, in general, the closest tag to the field of view will be read.

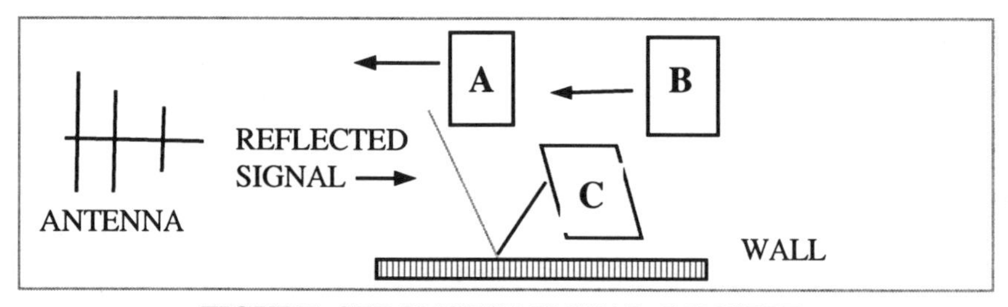

FIGURE: STRONGEST SIGNAL CONCEPT

One should also insure that the item with the tag is positioned or channeled to give it the best possible chance to be in the field of view. In the Figure: Strongest Signal Concept, Tag A is expected to be read, Tag B is blocked because, all things being equal, Tag B is farther away and should have a weaker signal. Tag C is reflecting a signal off the wall. It is possible to get a stronger signal from a reflected tag. It is also possible to find situations where Tag B is stronger than Tag A, and in such cases, Tag B might be read instead of Tag A. The solution to this is a quality vendor with a good design, and a high quality manufacturing process. Other actions are used in the design to control the location of the read, and the timing when the RF reader is turned on to accept the signal reflected from the tag.

ANTENNA CONCEPTS

There are antenna concepts that, if understood, can assist in the design of the system. The purpose of the antenna is to produce a signal that can be thought of as three-dimensional. Use the space or shape of a balloon to form an image of what is possible. The shape is not necessarily uniform, and there are no tracer bullets to show where the pattern is being generated. The patterns produced by certain antennas will provide a better or worse solution. While the signals shown below are two-dimensional, understand that the actual pattern is a three- dimensional pattern.

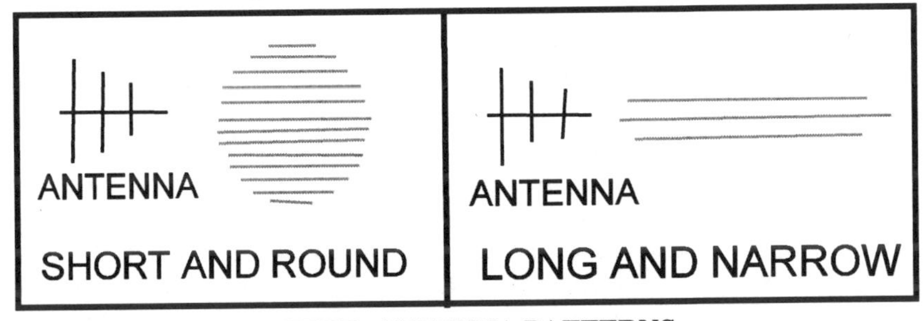

FIGURE: ANTENNA PATTERNS

The antenna patterns used for your application may need to be specifically for your application. The best operation of the system may depend on the antenna type and the patterns that are possible. The antenna you use may need to be controlled with a device that works with the antenna to assist in shaping the signal. These tools are known as attenuation devices.

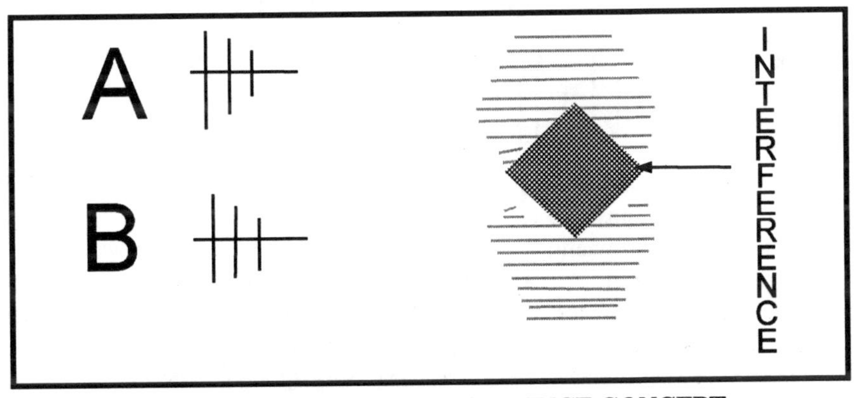

FIGURE: ANTENNA INTERFERENCE CONCEPT

When two antennas are in close proximity to each other, the signals can have the effect of canceling each other out at certain points. The interference is shown in a conceptual way above. The idea here is that if a particular pattern is needed to provide the plane for reading the transponder, then it is possible that the antennas will interfere with each other. This brings us to another concept about the pattern.

FIGURE: ANTENNA BACK-PATTERNS CONCEPT

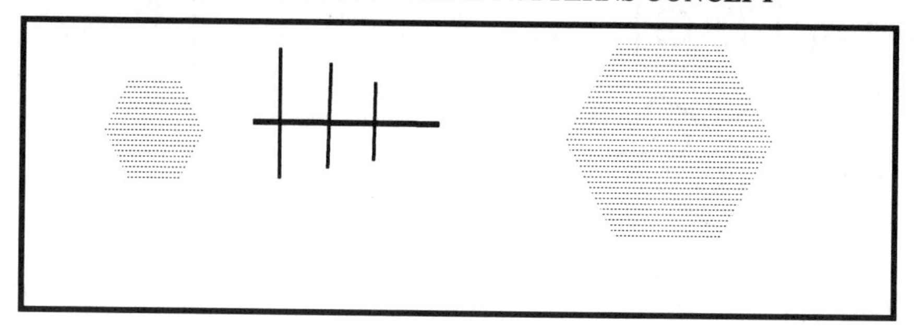

This concept shows that a given antenna may project a signal in both directions. The strengths of these signals should also be shown, for more completeness of detail, but for simplicity, the signal is often shown as only a foreword signal. If antennas are placed back-to-back, the signals of the rear of the antenna may interfere and render the system inoperable. Also, if a tag is very near the rear of the antenna, it may be read instead of a tag which is far from the of the antenna.

FIGURE: SHADOWING CONCEPT

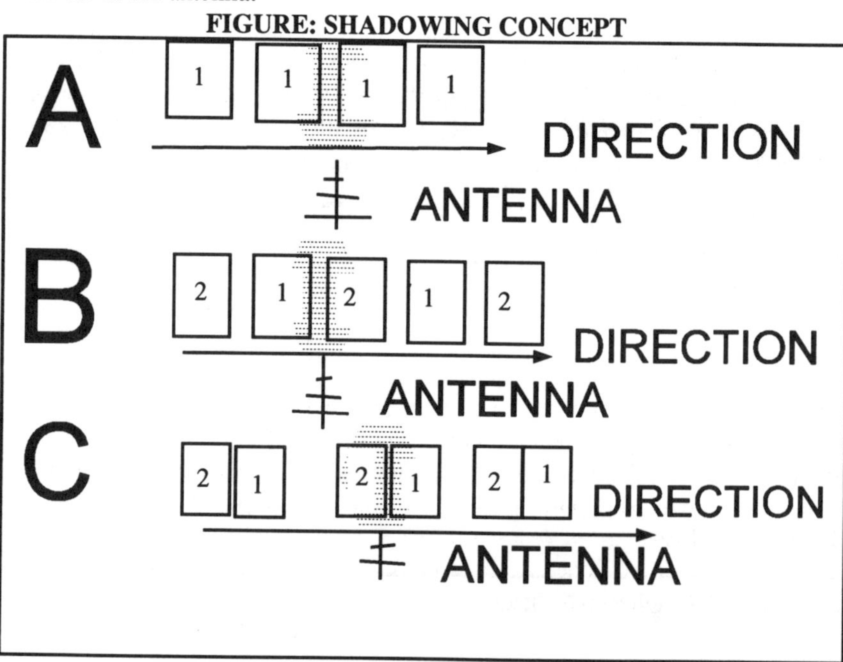

The shadowing concept is indicated on page 78. In situation A, Tag 1 is traveling past an antenna and the system reads the antenna as the tag crosses through the antenna pattern. In B, the concept shows two Tags, 1 and 2, passing the antenna and being read successfully. In C, the Tags, 1 and 2, are so close that one tag may be read and the other not read.

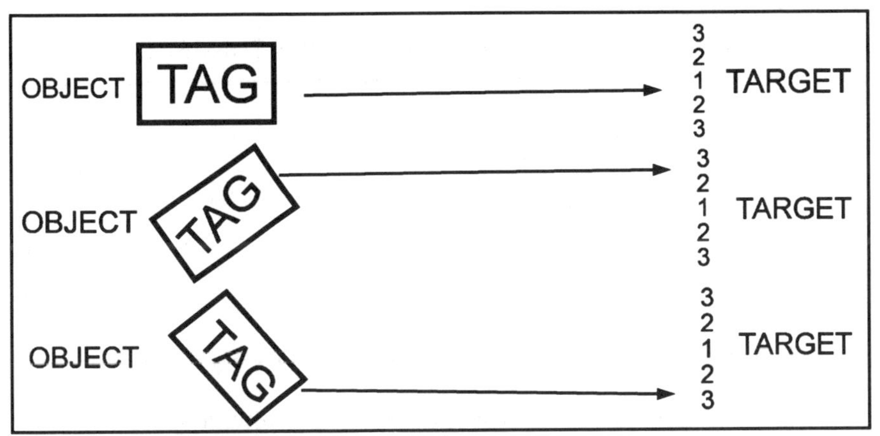

FIGURE: TAG/ANTENNA ALIGNMENT

The concept of tag and reader alignment should be considered. If the target is stationary, as shown above, then one might think the process normally has a tag pass by the stationary target. If there is a variable presentation path, the alignment may not be at optimum position. If the alignment of the vehicle is controlled, then there may be a problem caused by the vehicle swinging and swaying on the springs from uneven roadways, or the normal pitch and roll of uneven high speed travel. If there are tolerances for placement, then the extremes of the placement should be considered, with the all of the possible conditions.

For example, a tag could be mounted with a slight orientation pointing downward rather than perpendicular to the target. This variation combined with a target antenna ever so slightly pointed upward, and a car holding the tag rocking downward, could result in a missed tag read.

CHOKE POINTS

Radio frequency identification systems are used with a concept called choke points for producing information consisting of the ID and location, with time and date stamps. Choke points are fixed locations where some control point or observation point has importance to the end user. Examples of choke points are doorways, hallways, gates, lanes, roadways, ramps, and so forth. The idea is that something or someone is going to pass by the choke point, and this fact is important information to the user.

REDUNDANT READS

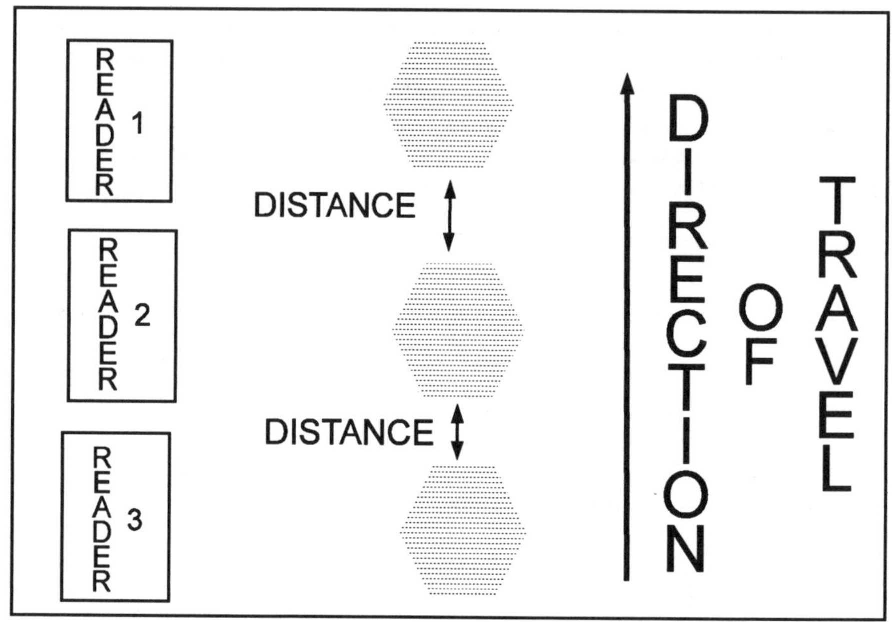

FIGURE: REDUNDANT READERS CONCEPT

There are times when it is so important to read the transponders as they pass by the system, that multiple readers will be placed in relatively the same place, but yet sufficiently far apart to prevent interference, thus insuring the highest probability of a successful read. The distance between the signals is a critical factor, so that the user can insure that there is no interference. Yet, the antennas need to be close enough to insure the same population of tags are being presented to all reader systems. The results will be combined and analyzed to produce an output that should be more reliable than the single reader.

FIXED-POINT READER SYSTEMS

The concept of a fixed point reader system is one in which the reader system is stationary. A transponder is affixed to an object, and the reader will interrogate the transponder as it passes by the reader location. The readers could be installed at, in-gates and out-gates, road sides, toll lanes and doorways, etc. As transponders pass by, the reader identifies them.

A clarification point here is that often a reader is stationary, but the more important point is that the antenna is stationary. Multiple antennas can be attached to a reader and each one, in this fixed point concept, is fixed to a location and covers an area of a designated choke point.

FIXED-STATION TRANSPONDERS

This concept is to have the transponder permanently affixed to a wall, post or some other stationary point, and to have readers pass by the transponder to read the ID information. This concept can be used to determine parking slots, mile markers, station location, bus stops, loading ramps and other locations. Readers can be mounted on tow trucks, buses, etc. This approach is best when there are few vehicles and many locations of interest.

WIDE-AREA APPLICATIONS

The concept of wide-area comes into play as a technique that will allow for identification of all transponders in a given geographic space, at essentially the same time. Transponders enclosed in a container could be identified. All transponders in a warehouse could be identified. The notion of wide-area has varying frequencies within a range, broadcasting in an area. Transponders lock on to a given signal, to reply to the reader and then go to sleep while other transponders reply. Inventory applications can use this concept.

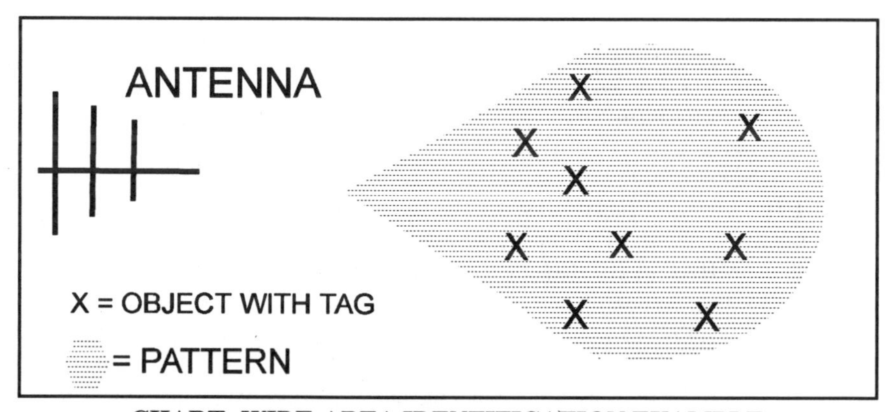

CHART: WIDE-AREA IDENTIFICATION EXAMPLE

HYBRID APPLICATIONS CONCEPTS

Hybrid applications are those that include the automatic equipment identification technology in combination with other technologies to produce information beyond a simple identification. This is a very powerful application technique, and is one that brings significant advantages to the data collection industry. The power of automatically and reliably collecting the ID along with other information, and making the data available to the user community is revolutionary.

This concept will directly save millions of dollars and make cost avoidance available for the enterprising user community. The primary reason for this is that the applications will provide another level of operational data for managers to use to make more informed decisions. In other cases customer service levels can be improved for a reasonable cost or investment. Productivity improvements will also contribute. Because the data can be collected automatically, accurately, and easily, at a fraction of the cost that manual processes required in the past, a whole new set of applications can be implemented. Processes can be re-engineered and additional savings incurred.

What is a hybrid ID application? The term hybrid means bred from two different species or of mixed origins. In the case of hybrid ID application, there are multiple data sources, that when combined into an integrated system, provide powerful information to the user community.

For example, shock detection could be combined with a radio frequency identification device, and both of these could be associated with a container. The end user, who normally would have an ID and a shock detection device for the container, would now have an ID/shock detection device for the container. The advantage is that the manual process, that is so often fraught with error to associate the two data points, is eliminated.

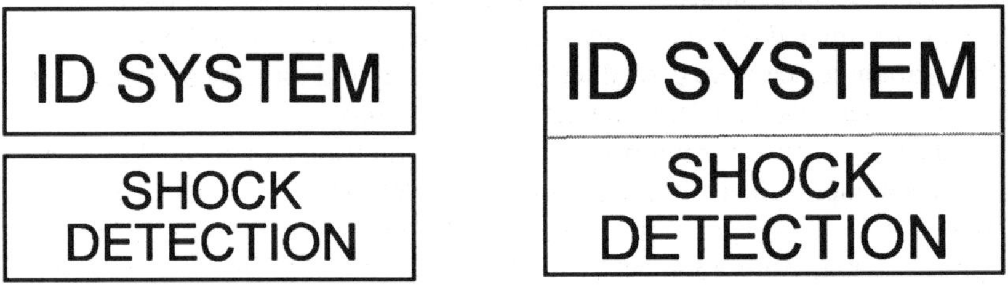

ID SYSTEM

SHOCK DETECTION

ID SYSTEM

SHOCK DETECTION

CHART: HYBRID ID APPLICATION EXAMPLE

RFID APPLICATIONS

The uses for RF/ID may seem evident, but it is worthwhile to review these basic concepts. Things can be identified at choke points, at fixed locations, and in larger areas. We can combine the identification process with other data collection technologies, and form hybrid applications. These also can be integrated into our processes to provide the management and operations personnel information upon which to take action. Fundamentally, the RF/ID system can be used in several application concepts.

1.Gate 2.Inventory 3.Sorting 4.Security 5.Closed Loop

GATE CONCEPT

Everyone is familiar with gates. Gates are everywhere. Fundamentally, a gate is a door or a valve or an opening in a wall or fence. There is a notion of a confined space or choke point in the gate concept. An entrance to a building is a good example. Parking lots have gates. As you start searching for gates, you begin to see them, and typically you find more than you expect. The idea is that you can begin to identify things at gates, which are good choke points. With ID systems the question of what passed this point, and when, can be answered. Chapter ten is dedicated to Multi-Media Gate Systems.

INVENTORY

An inventory is a collection or list of items. In this application, what is in the yard is identified. If RF/ID is employed, the question of what items are now in the store could be answered in a few moments. What trailers are located in this section of the parking lot, could also be answered in a relatively short fashion. With hybrid location and ID systems, more detailed lists can be obtained. For example, what parking lot holds this particular automobile? It is possible to identify the transponders in a specific area, and collect location data.

If the transponders are affixed to containers in an intermodal yard, and the containers are parked, they become stationary or fixed location containers. If a reader were to be attached to a pickup truck, as an example, then the truck could drive up and down the lanes, and the reader could read the transponder. If a location device were placed on the truck, a hybrid location system could be used to generate a list of transponders and locations.

Location systems for interrogating the RF transponders, located in an area, have been approached from both concepts of RF/ID, moving the reader down rows of transponders, and using wide area RF to understand what is located in a general area. Both approaches might use the hybrid technique to enhance system design.

SORTING CONCEPTS

To sort is to arrange by sequence or class. The concept of sorting is used all of the time. Destination zip codes come to mind. If packages had RF/ID's with destination zip codes, then automatic sorting might be possible without the effort to orient the package with the label facing up. Low frequency technology could be used to read the label through most packages, without the orientation requirement. This application needs a way to get the items being sorted, separated from each other, so that diverting machines can channel items to the proper sorting channels or bins.

SECURITY CONCEPTS

To be secure is to be safe from loss or to be certain or sure. Identification systems can provide a means to be certain that the identification code is authorized, and the equipment is confirmed as authorized to enter or leave a gate. Identification of the drivers and the equipment helps to validate the transaction.

CLOSED LOOP

When something travels around in a circle or back and forth to the same location it can be thought of as traveling through a closed loop. This concept is a good initiation when starting out with the RF/ID technology. You can limit the exposure of having to put tags on the entire fleet, before you get the opportunity to view the application in your own working environment. If, for example you have tractors pulling loads from Coral Springs, Florida, to Cleveland, Ohio, and then to Detroit, Michigan, and back to Florida you could think of this as a closed loop. See below.

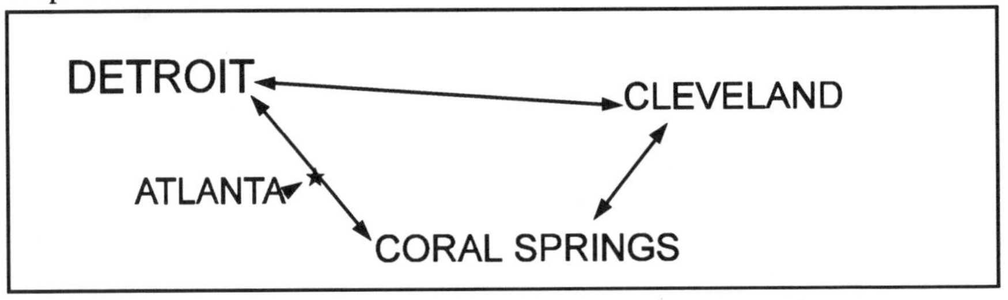

FIGURE: CLOSED LOOP EXAMPLE

In the example shown, the three cities of Coral Springs, Cleveland, and Detroit are in a closed loop. RF reader stations could be placed at the gates of the terminals at each of these three locations. When a tractor, that travels among these three locations, enters a gate, a transaction can be logged showing the date and time, and ID from the transponder affixed to the tractor. The point at Atlanta is shown because, in this example, a refueling station may be needed, and a reader could be placed at a designated station in Atlanta.

The idea of a closed loop applies if part of the fleet travels to the same destinations, even part of the time. In the above example, tractors could also just travel between Detroit and Cleveland, and the gate systems could record the events.

COMMON "GOTCHA'S"

While the basic ideas are sound for RF/ID systems, there are a few things to look out for in the process of using the transponders for the identification, and the reader equipment to read them. For one thing, there are no tracer bullets, so it might be difficult to determine where you are pointing the antenna system. You get the idea. This is not meant to be an all encompassing, fully exhaustive list, but it should give you an idea of things to watch out for in the process. Some of these have been taken care of in product enhancements, and others are things that, over the past six years or so, have come up in discussions.

1. The tag ID in the tag is not the intended ID. Verification techniques and double checking can help.

2. The tag ID on the equipment is not the one for the equipment. Again a verification process is needed and system routines to check for the occurrences can help.

3. Tag data is modified via an external source. Certain signals and signal strengths can damage certain tags, after the identification has been encoded on the tag.

4. Portable reader systems might affect the readability of the tag. This was an early problem believed to be solved.

5. The tag is read by another reader system or the signal bounces off a metal building.

6. The tag is read intermittently by another reader, caused by reflections from trailers and containers somewhat randomly parked in a nearby parking lot.

7. Two antenna signals cancel each other out at certain points.

8. One antenna reads a tag in line of site but across several lanes.

9. The antenna position severely restricts the ability for highly reliable reads. Tags may be mounted on equipment with a desired read position at or close to limit of specification. (Example: when the antenna's height is restricted.)

10. Tag separation was insufficient to detect a second tag.

"GOTCHA " REMEDIES

Certainly, these problems must be categories for exceptional situations. The fact is that the history of RF/ID is not long enough, and all conditions may not have been introduced to the systems. Certainly, as the standards organizations provide test criteria, and manufacturers produce stress tests, improvements will continue. The design should improve, and more features will be forthcoming . There are ways to check for these conditions, and to monitor the equipment, so that error rates can be minimized. In addition, there are some tools available to assist us in controlling the signals.

A. Presence detectors can be used to turn on the reader system, only when the equipment or object with a transponder is in the area.

B. Attenuation devices can be used to control the antenna pattern and to change the shape of the signal.

C. Redundant systems can be used to compare the results.

D. Multiplexed antennas can be used to control the signals that are active.

There are also ways to test the reader sites periodically, to insure that the antennas are focused on the desired areas. In addition, there are ways to analyze the systems data to detect the degrading operation of the system site.

PACKAGING

There are a number of packaging options that are used in the industry as provided by the RF vendors. Options are still being created to fit customer needs. Primarily, there has been an approach that fostered the integration of the hardware to fit the customer application. Some manufacturers are making units that are rich in environmental conditioning. The primary components, if not all, are packaged to meet the outdoor environment. Often there is a need to customize the equipment to include the overall components of the system. The economics are such that there may already be enclosures and other space available. In these circumstances, the work effort is the integration of the RF devices into the existing environment.

Some systems require that an office environment support the computer system, and that the antenna and other equipment are mounted in a harsh environment. The diagrams show this factor. If the system is not readied for the harsh environment, then additional considerations are required and the functions for power, surge protection, heat, and cooling will need to be added. The rail example shows the concept.

Consideration must be given to the features required to keep the system running and the features needed to communicate with the often remotely situated RF reader system.

It is often necessary to include multiple communication links to the system. In some cases, these are for backup and recovery purposes, and in other cases, for support of multiple operating environments. Depending on the system design and the user availability required, a phone line may be required for the reader to send out information. Another line may be required for the control point to communicate with the reader while it is operating. In other cases, a data radio or data modem capability will provide communication with a base station.

Other system capabilities require features including auxiliary power or battery backup. Fans may need to be installed to control hot or cold air flow. A heat exchanger may be required. Various digital input and output points may be needed for the collection of the information, and to invoke a desired action or control. Stabilizers may be needed for the special housing required to hold the components. Special grounding packages may also be required. There are a number of options to consider in the design process.

Some manufacturers are providing simple reader packages that include the reader, RF module, antenna and some memory and I/O ports in a simple to install package. These reader systems have standard interface capability to computers for system control purposes.

SUMMARY

With these concepts in mind, it is now possible to design end-user applications for the maximum benefit. These concepts and specific applications will be brought out in the chapters on applications. The RF/ID systems work. There are some cautions.

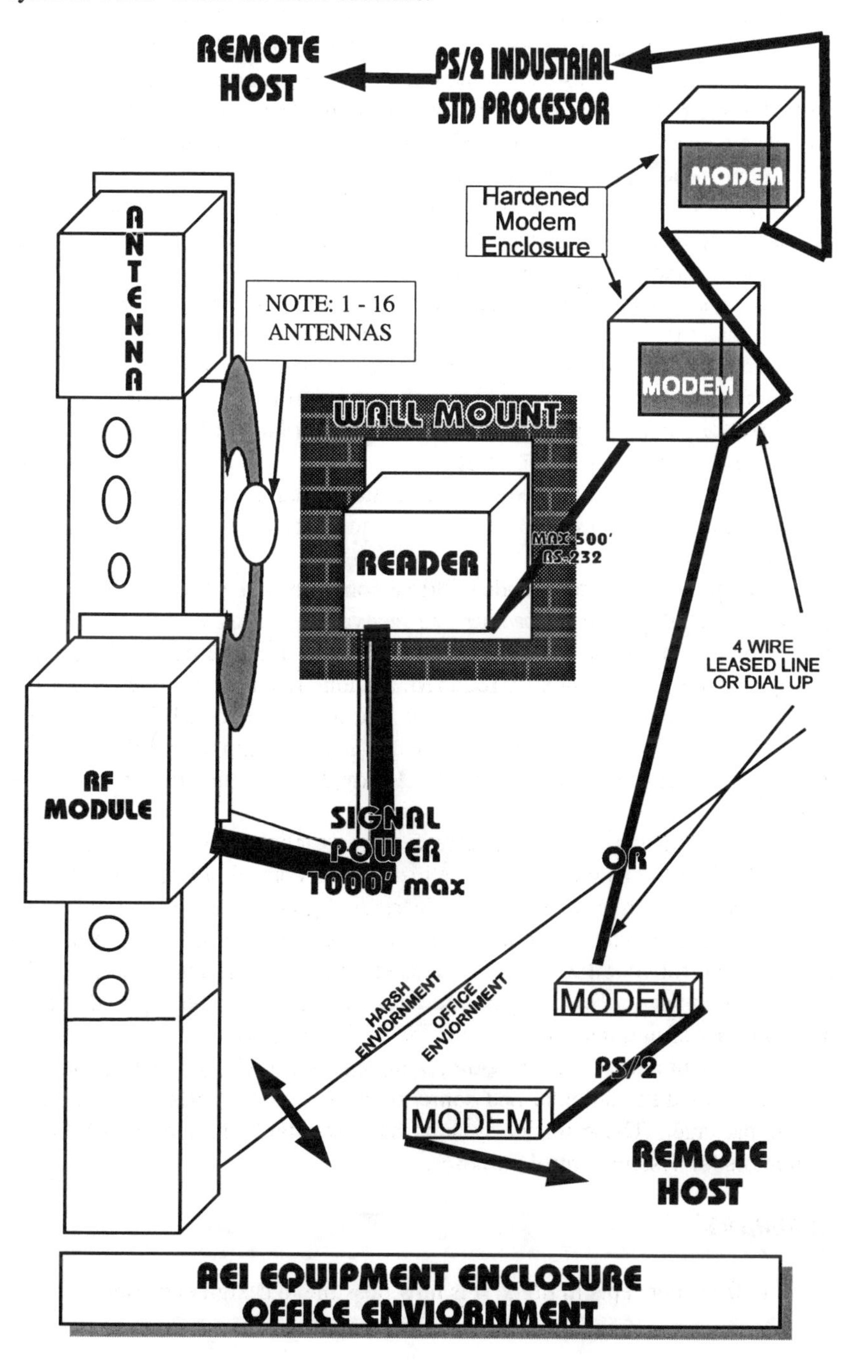

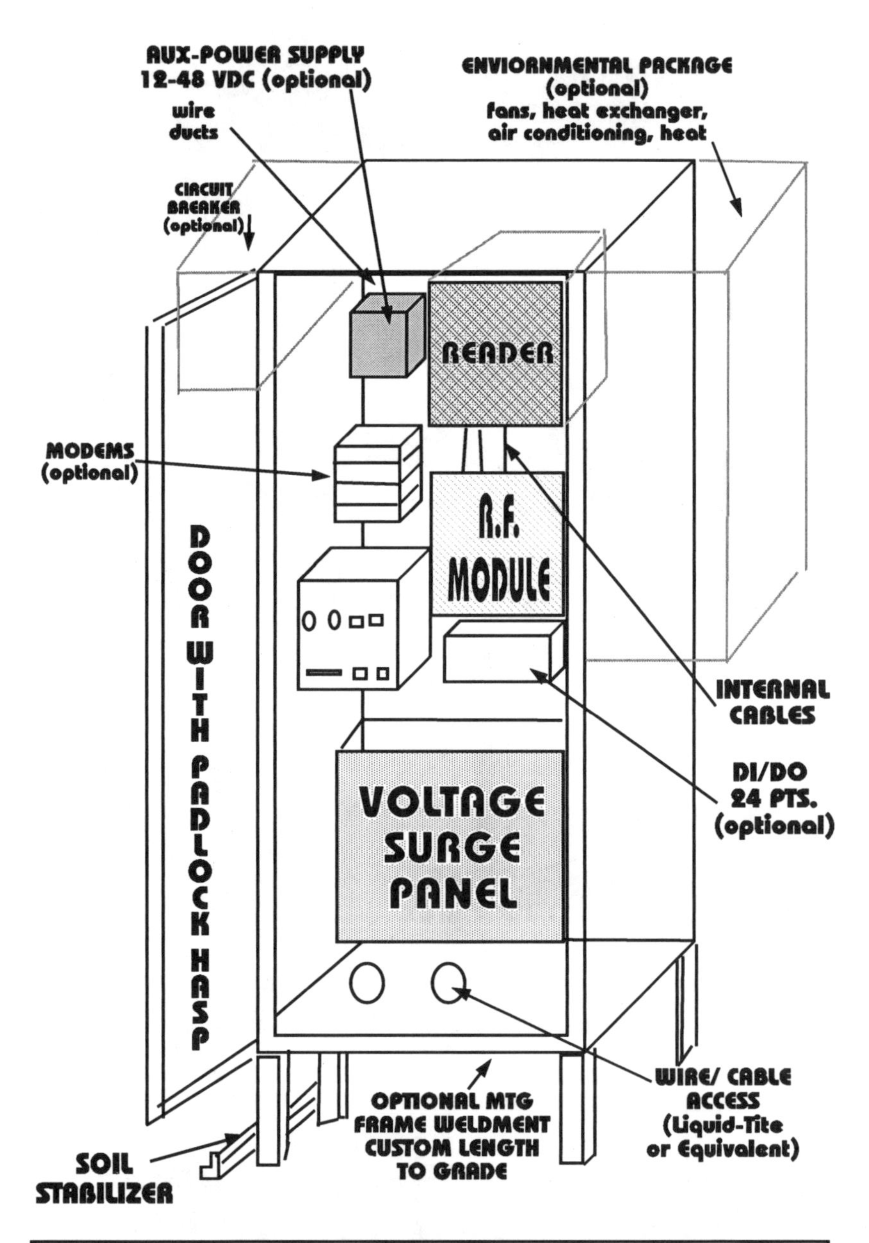

AEI EQUIPMENT ENCLOSURE RAIL TRACKSIDE

CHAPTER X

GATE AND OTHER USEFUL TECHNOLOGIES

Technology has been advancing rapidly. Our PHD's are now aided by vast computer power to help them develop the technology to change our world. The pace is fast. Some of the new technology fits well in the gate environment, and it is now possible to totally automate the gate processing. In addition, location and communication technology has taken a great leap forward. These technologies are explored with reference to how the RF/ID technology will fit in hybrid applications.

GATES

Gates are everywhere! It is a human phenomenon to use gates. There are gates to keep things in, and gates to keep things out, and then there are gates to control the in and out activity. Some gates have physical obstructions to prevent passage, and others are open and seem happy to merely account for the passage. Some gates have attendants and others do not. The next time you drive to work, not only be cautious in your driving, but observe the gates that are along the way.

There are tremendous responsibilities in operating the gate. Responsibility for permitting an expensive piece of equipment and even more expensive cargoes to leave the premises is one. Accepting the responsibility for receiving the equipment and shipments is another. Permitting a bus to leave the gate prior to safety inspections can cause significant problems that hit the 11:00 PM news. But for most of us, it is an opportunity to capture operational data and for transacting business. We enter parking lots through gates, pay tolls at gates, and surround our homes with gates. We go to work through gates, go to sporting events with gates for the cars and gates for the fans. Gates are everywhere!

FUNCTIONS AT GATES

There are many different functions performed or that can possibly be performed at the gate. Certainly, there can be a verification, a security function, and a control function. The degree to which each of these functions are required, and the degree of confidence required for the company or organization will dictate just what technologies should be used to facilitate the process. There may be a function to keep out casual observers or to protect with maximum security. Because a gate is a choke point in the operation, it is a good place to gather data about the operation and to checkpoint key items of interest. The gate is a good place to capture operational data while in the course of normal operations.

There are some key objectives that should be the focus while we are at the gate. The objective is not to slow the transaction through the gate, it is to speed the process through the gate. For the mathematically inclined, the gate is a single service queue. If you slow the process through the gate, you may get long lines. In fact, if you slow the process, it is guaranteed to cause long lines, usually in both directions. That equates to long wait times, or more gates to handle the peaks. It might even cause bi-directional gate lanes assuring heavy traffic before the game, when the gate is an in- gate; and after the game, when the gate is an out- gate.

It is not acceptable to speed the process through the gate, if you forget to process the necessary transactions. The objective is to process every required transaction, and do it in the least amount of time possible. The first thing you may need to do is eliminate anything that may be considered as not required. After that, you will need to look at technology to see what might be done to modify the processes, or to totally re-engineer the processes. This is where technology can assist in a very big way.

LOW FUNCTION GATES

The concept is that not just anyone can pass this way. If you don't have a ticket, you can't get in. If you have the ticket, which in this case might be an identification badge, then you can open the gate and enter. Whoever has the badge can gain access. This is why the RF/ID system is perfect for this application. The transponder is placed on the vehicle, or in the vehicle, and then if the identification code is authorized, the gate will open. The functions of this gate are relatively simple, but very effective. You would have to call a security guard or someone else if you do not have the transponder.

MULTI FUNCTION GATES

Obviously, there are gates where more functions need to be performed than a confirmation of the possession of an RF tag. The processes needed at these gates require significantly more information. There is more to be considered.

A. Security
B. Inspection
C. Control
D. Communication
E. Data Collection

Each one of these areas have multiple options for success. The options might be considered added features. The functions within the process can coexist and the information from the process confirms a point, or adds information to the total. The probability of success is increased if a security guard confirms the company ID, drivers license or work orders and visually recognizes the driver and so on. All of these factors insure that the proper person is identified. The factors add up to confirm the identity. If the driver

PHOTOS BY GERDEMAN

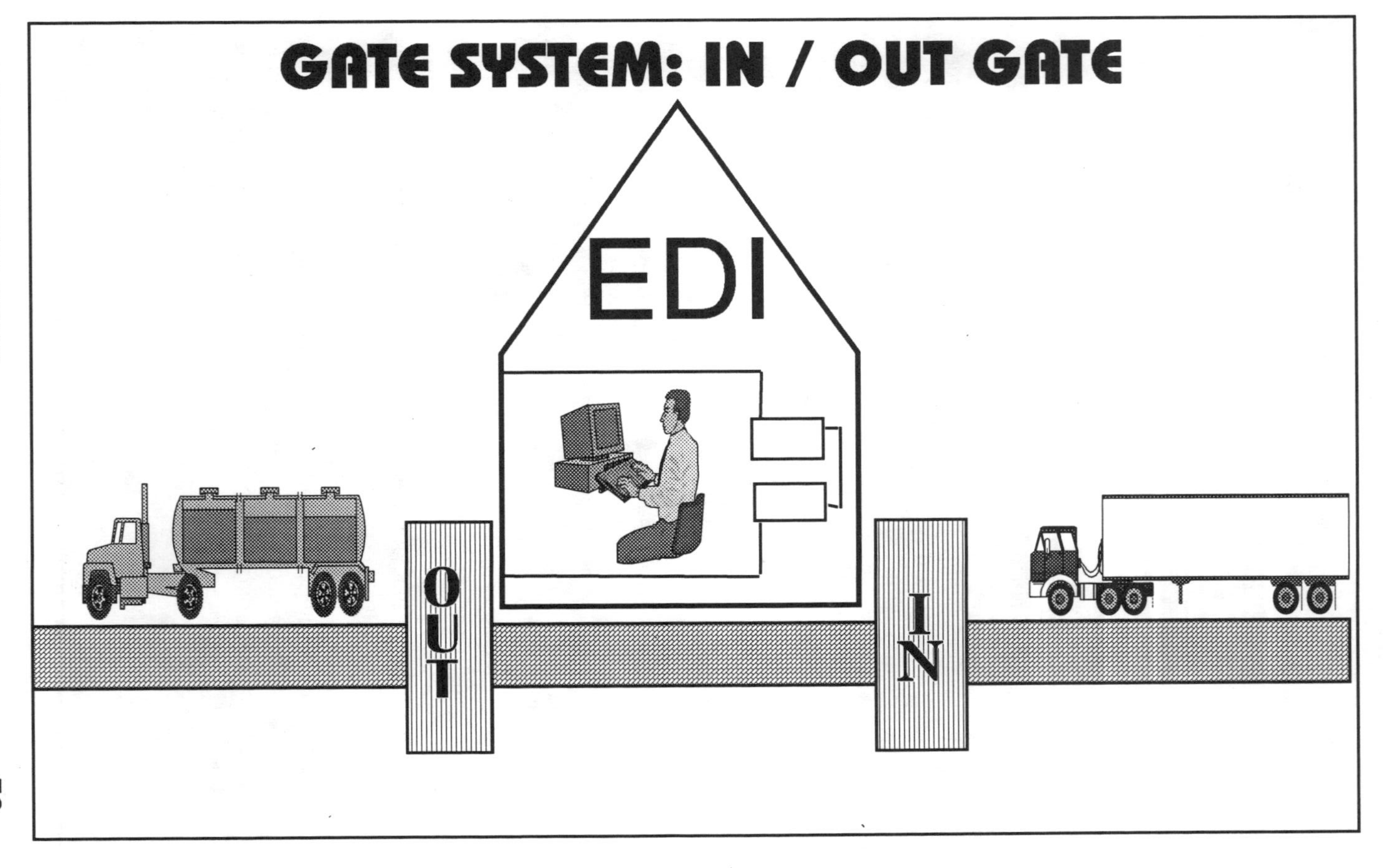
GATE SYSTEM: IN / OUT GATE
EDI
OUT
IN

only had a company ID without a picture, the security guard may suspect something.

SECURITY

To be secure is to be safe against loss, attack, or escape. While there are a number of considerations that the experts can provide, there are a number of things that RF/ID can do to assist. There are also some other technologies that are emerging to provide a secure gate environment. The system is expecting a certain set of identification codes at a certain gate. If a code shows up at a gate, you can confirm the ID, and open the gate. In intermodal and trucking applications there may be multiple pieces of equipment that will enter the gate area. For example, a triple rig can have a tractor, three containers, three chassis and a couple of dollies. All nine pieces can be identified.

The driver can also hold an ID card with RF/ID features. The driver ID could also be read into the system automatically. A personal identification number, or PIN, could be used as well.

Security options might also include using the read/ write capability of the transponders. In these cases there could be designated fields from inspection systems. Shipment information could be written into the tag at the origin point, and transmitted, via EDI, to the destination and intermediate locations. It could even be possible to have the transponder hold an image of the driver. There are a number of options to explore that could revolutionize the security use of the RF transponder.

OTHER SECURITY TECHNOLOGIES

Some security technologies include the use of multiple techniques to identify the people and the equipment that enter and depart a gate. There has been significant progress made in voice recognition. Pattern recognition has also increased in accuracy for automatic picture recognition.

The problem with using ID cards at a highly secured gate is that they can be duplicated or stolen or both. If a thief has been recognized from prior activities and the guard knows and remembers the individual, then a positive action can be invoked. If the process needs to be automated, then there needs to be a way to capture additional data in the normal process of the gate.

PATTERN RECOGNITION

Recently, pattern recognition technology has taken off. There are cards that can be put into personal computers that can accept images from relatively normal function cameras. These cards will grab frames of pictures, analyze, enhance, and manipulate the data, to provide all kinds of information to be analyzed. Keep in mind that to analyze a photograph, there is a significant demand on resources. These resources include memory, computer processor speed, and storage, and of course, card slots.

Now the cards have processors and memory of their own. When you use these highly specialized cards, the functions can be performed very fast and often at sub- second speeds. There are many ways to capture the images, and while the implication above is that a relatively unsophisticated camera can be used, there are advances in cameras that can assist and enhance the image as well. The speeds of the processors are very fast and are moderately priced. A large amount of memory is available for a relatively low price, and the same is true for storage. Applications that were not affordable in the past, may have crossed over to the affordable column. So, what else is needed to complete the pattern recognition?

The software that manipulates the data has taken a major step forward. Using decision support techniques and the artificial intelligence capabilities, certain pattern recognition can be accomplished. The notion of using neural network techniques is coming quickly to the market. Some software is available to use a feature extraction technique to identify individuals and other objects. The facial analysis is so advanced, that even with or without facial hair, and some amount of makeup, individuals can be recognized.

SPEECH - VOICE RECOGNITION

Now, imagine that the person we want to capture at the gate, has a legitimate RF/ID tag, and has had major reconstructive surgery on his face. One expert thinks that feature extraction would still work, but let's say that it will not. As we speak to the guy at the gate we can capture his voice. As we get more and more input from the driver, there are ways to use the neural network to recognize the individual, even with an attempt at disguising the voice. This will need more work but again the basic ideas are being worked on and developed for system users.

INSPECTION

To inspect is to look over carefully, to examine officially. At a gate there are some particular things to inspect. Primarily, are the vehicles in good shape physically. Has any damage been done to the equipment while in the gate, or has there been damage while outside the gate. So, before and after images are important and an automatic identification is required. The RF/ID system can identify the equipment. This identification is a significant event in the process. If the identification is from a key entry then there are levels of errors introduced into the system. If the identification that is written on the equipment is used, there is another level of error, the letters may be faded, erased or transposed, etc. Automatic equipment identification provides us with confirming information that has been written on the transponder.

The identification code is used to index the records stored. With reliable and repeatable ID tags, and the technology to take photographs of the equipment, key inspection processes can be automated.

Photos of the sides of the equipment can be captured. Line scanners and other types of cameras can be used to capture the images. The neural network can do image processing to accomplish pattern recognition and establish patterns. Before and after photo comparisons can be accomplished automatically, even when different cameras at different gates are used to capture the images. Photos of before and after can be stored and comparisons can be returned, printed and analyzed on demand. If there are disputes, which tend to come later in time, often days or even weeks later, a photo history can be produced to help with the settlement.

The images can be stored in our computer systems. The storage requirements can be very significant. 60,000 bytes per image is quoted from some sources with good compaction routines. As the number of image captures rises, there may be difficulty transmitting these images to host systems. A large communications pipe is needed. Fortunately, storage capacities are growing by leaps and bounds. So is the price performance factors. Sometimes the image is stored locally, and then, only the needed images are transmitted to a central site.

If you expand the thinking a little further, you can even do some safety oriented inspections automatically. These are inspections that look for wear and tear, things that may easily be missed by the naked eye. If there is a way to control viewing the under side of the equipment, then cameras can be placed in positions to capture the necessary images for further analysis. Feature extraction and processing techniques can assist in determining that a part is worn to a marginal situation. The tires can be examined for flats or tread. Even a pressure indication might be possible. Other inspections like break wear is possible. This is a much more complicated process, and the processing techniques will need to continue to improve, but it is a very good possibility.

CONTROL

Control is power, authority and the means to regulate. If something is permitted to enter the gate, and the criteria is met, then passage can be enacted. In the case of our radio frequency identification gate, the tag would be read and if the computer matched the ID to a valid table, then the system could open the gate. If, however, the tag was invalid, then additional options, like do nothing, or notify the security control center, can be acted upon.

The control process may mean a check on the paper work to see that it is in order. There may be identification and inspection data and then the control may need to insure that the bills of lading and weight and fuel are justified for passage. Is the equipment insured? In other words are the laws, regulations, and other requirements being met. Then is there anything that the company needs to know or collect prior to entrance or exit. Are the import/export papers in order? These are things that the company is responsible for and must be in control of, and for which the company could be subject to significant penalties if found not in compliance.

While this part of the process is often called bureaucratic, it is a very important step. RF/ID systems can assist in the process by making several of the steps more reliable from the identification process. If the paper work could be kept on the transponder through a set of processes, security codes, encryption and other design characteristics, then the system could write the data onto the tag at specific stations in the flow of the processes.

This technology may be one of the few that could get us to a point of a truly paperless logistics flow. So how would this help? If the paperwork was not completed at a checkpoint, lets say the weight was missing, then an order could be given to go get weighed. At the weigh station, the ID could be recorded, and the weight information sent to the host computer. At the same time the weight field in the transponder could be updated, if the weigh station is authorized to do so. When the vehicle reaches the checkpoint this time, all systems will be go.

Other technologies that can assist in this process of control might be the image capture of the documents, EDI, and bar code. Signature capture methods will also assist in the flow through the gate. These techniques and data capture technologies can be sent to the gate and be made available to the gate keepers through a communications network.

COMMUNICATION

Communication transports the information to all the parties and systems. The data must be accurate, and available on a timely basis. We have seen how RF/ID systems improve accuracy and timeliness. The data must get into the control system, so that the management and clerks can have immediate access to the data in the system. Telephone communications may be essential. Some systems designers need satellite communications. Perhaps cellular is the right capability. Dial up capability, leased or dedicated lines or both may be required. It is essential to have a good communications capability from the gate to the back office system, from the gate to the RF/ID system, and also from the gate to the drivers.

Speaker or intercom systems can be used to communicate with the driver. Speech recognition systems can even determine the language of the driver, and automatically provide instructions in the driver's native tongue. Signs can also be used to direct the driver through the gate, and to provide directions to the location of the drop or pick up point.

VIDEO

STOP ACTION

LINE SCANNERS

CCTV

FULL MOTION

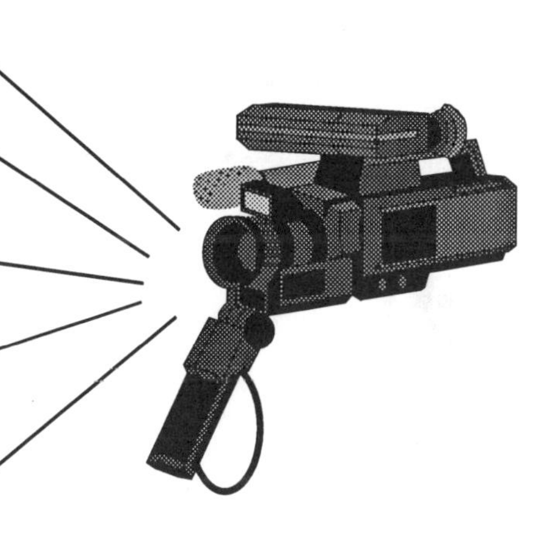

* SECURITY ACCESS
* REVENUE COLLECTION
* VERIFICATION
* VEHICLE CLASSIFICATION
* ENFORCEMENT
* INCIDENT DETECTION

DIRECTIONAL SIGNS

DATA COLLECTION

If the best time to collect information is during the normal course of conducting business, then the gate becomes an excellent place to collect data other than that which is needed for control purposes. One of the best examples is to use the gate as a place to collect maintenance information. The RF/ID system using a dynamic tag, can provide information for the maintenance department. The recording of mileage, fuel level, high rpm's, oil viscosity and coolant status may all be possible.

If the weight was determined by a service, and written to the transponder then the gate can be a good place to capture the weight for our systems. Likewise, this might be a good place to collect government clearances. So, while the process through a gate has many functions, it is also an excellent time to collect data to feed other systems.

GATE APPLICATIONS

There are a number of gate applications that come to mind when you begin to think of the gates involved in your own life.

AIR FREIGHT	GOVERNMENT FACILITIES
AIRPORT PARKING	HOSPITAL PARKING
CONSTRUCTION GATES	HOTEL PARKING
COUNTRY CLUB ENTRANCE	INTERMODEL GATES
DELIVERY	LIMO PARKING
DOCTOR S' PARKING	MOTOR FREIGHT GATES
DOORWAYS	PARKING LOTS
EMPLOYEE PARKING	PARKS AND RECREATION
EQUIPMENT TRACKING	PUBLIC PARKING
EXECUTIVE PARKING	RAILYARD GATES
FREQUENT FLYER	SERVICE FLEETS
TAXI PARKING	STUDENT PARKING
TOLL GATES	VIP PARKING
WALKWAYS	REPAIR SERVICE

CHART: GATE APPLICATIONS

Some view gates as all the same. In fact, there are many types, with different levels of functions. Certainly the functions implied above add to the complexity of the gate.

GATE TYPES WITH FUNCTIONS

TYPE	FUNCTION(S)
LOW FUNCTION SECURITY	This gate is a barrier. It is either open or closed. There is a lock that can be opened with a key.
UN-ATTENDED	This gate does not have an attendant in the guard-house. An identification is needed to enter. RF/ID can make this relatively hands free.
FEE GATE	These gates use coin collection, or token collection to open the gate. It is used for toll and parking collections. RF/ID can be part of this system.
ATTENDED GATE	This gate has an attendant and is usually a cluster of gates. A mixture of automatic and other gates are in the cluster. The advantage is that when something goes wrong, the attendant can assist in the corrective measures. RF/ID can assist here as well.
UN-ATTENDED WITH REPORT	This is similar to the FEE or UN-ATTENDED gate, but if an unauthorized person passes through, the system will capture the event and report it. Violation enforcement systems use this concept in the toll collection environment. RF/ID is a good fit here.
INDUSTRIAL	These gates imply that not only is there security, but there is a process to control what comes in and out of the gate. RF/ID works well here.

TABLE: GATE TYPES WITH FUNCTIONS

PARKING SYSTEMS

RF/ID can provide a significant change to the management of parking facilities. In fact, there may be a way to revolutionize the parking systems and improve customer service at the same time. When multiple lots are involved, available versus closed status can be maintained. This however is not new, but to dynamically change the price and use of the parking lot may be. For example if a parking subscriber for a city lot parks in the lot on a weekly basis, an RF/ID tag can be given to the subscriber for his car. The same tag might be used at airport parking lots and other parking locations throughout the city. A fee for the preferred lot is charged based on monthly use, and pay as you go fees are applied to the other lots. The user of the system just drives through the gate as it automatically opens. Often there are special event parking rates for city

FIGURE: APPLICATION CONCEPTS

parking lots. These come into effect on Saturdays and Sundays when the work week is completed. They might also apply to evenings as well. If the special event, such as a ball game or concert is scheduled, then the rate could change from a monthly rate to a special event rate, if the subscriber asked for the privilege.

Other high demand systems could be employed. Student parking can be a problem at universities and colleges. On- campus parking is generally reserved for the faculty and administrative help. However, parking on campus at night may be available for night time students and library access, etc. When the appropriate time comes, the system will permit students to use the lot. In the morning the system resets for faculty only. If a student over stays their welcome, there is a fee assessed, and the system will not permit entry into the parking lot until payment is received.

Parking lot gate systems can provide some interesting application opportunities for the owners and managers of the lots. The customer gets additional service and convenience as well. Coming soon there may be a significant change in the parking system at your local lot.

OTHER TECHNOLOGIES

When it comes to managing fleets of vehicles, and the crews associated with them, and there are many different equipment types to be concerned about, additional help may be needed. Throughout the various industry oriented chapters there have been references to three key technologies in addition to the radio frequency identification technology. These include:

1. Global Positioning Systems - GPS
2. Artificial Intelligence / Neural Networks
3. Data Radio Communications

GPS

Global Positioning System uses a constellation of 24 satellites that are specifically placed in orbit around the globe. The GPS system provides accurate position information to government and commercial users on a worldwide basis. The system is a direct ranging radio navigation system. The capabilities of this system to provide exact location data is very accurate. The government controls this system, and when it is employed for military use, centimeter accuracy is not out of the question. Commercial use is not as accurate, but for determining where something is located, this system is very good. To provide marine locations it is acceptable. Accuracy of a few meters is possible and there are vendors who report even better results.

To achieve accuracy in the commercial market, the concept of using a differential technique is employed. In this concept the satellites communicate with two GPS receivers. One receiver is on mobile equipment, such as a top-pick and the other is located at a base station. The base station is at a fixed location that has been surveyed to determine the exact location. As the satellites communicate the location to the base station a location calculation is made to determine an error factor for each satellite. The corrected factors are broadcast to the mobile units so that a differential location can be determined.

GPS is a very good as a technique to determine the location of objects or units. In fact, a three dimensional position and a velocity is provided by the system along with an accurate time. There is worldwide coverage for determining the location. The system works in all kinds of weather. The system needs a line of site with the satellites, so, if there is a cover such as trees or a roof of a building, then the system cannot work. City environments with huge skyscrapers produce caverns where multiple satellites cannot be contacted. So there are some drawbacks. If the line of site is clear this system can be extremely accurate. The use of the satellite system is free but there is a cost for the GPS unit. There is a long term government support commitment as well.

With the GPS system to provide the exact location of the equipment, and the RF/ID system to provide the identification of the equipment, accurate location data can be made available to the location data bases and the planning systems.

DATA RADIO COMMUNICATIONS

Hand held computers, on board computers, laptop computers, personal communication computers are all useful with identification systems. There is a full function general purpose system available to enhance the system capabilities for use in gate systems and other applications. There are more computers used in the power units of our fleets today than ever before. Communicating with these real time systems through the use of a general purpose computer will provide functions like real time status. Location systems like GPS can interface with these general purpose computers for transmission of location data to a central site or other host systems. RF/ID systems integrated into the same computer can provide additional reliable data.

The RF/ID may be used as a part or piece of equipment or even as the tool used in the process. There may be high frequency ID or low frequency ID applications or both to determine an accurate data entry of the necessary data. For on- board computers there may be the possibility of an antenna connection from the tractors to the trailers. Or this connection may be from the transponder to the container. A positive identification is made for analysis. With the combination of location data from the RF/ID system and the GPS system and the communications power of the computer there are a multitude of benefits to be gained from knowing that the specific equipment is located at this specific location.

With other inputs now available to be captured through the use of the computer located on the equipment, more detail equipment information can be

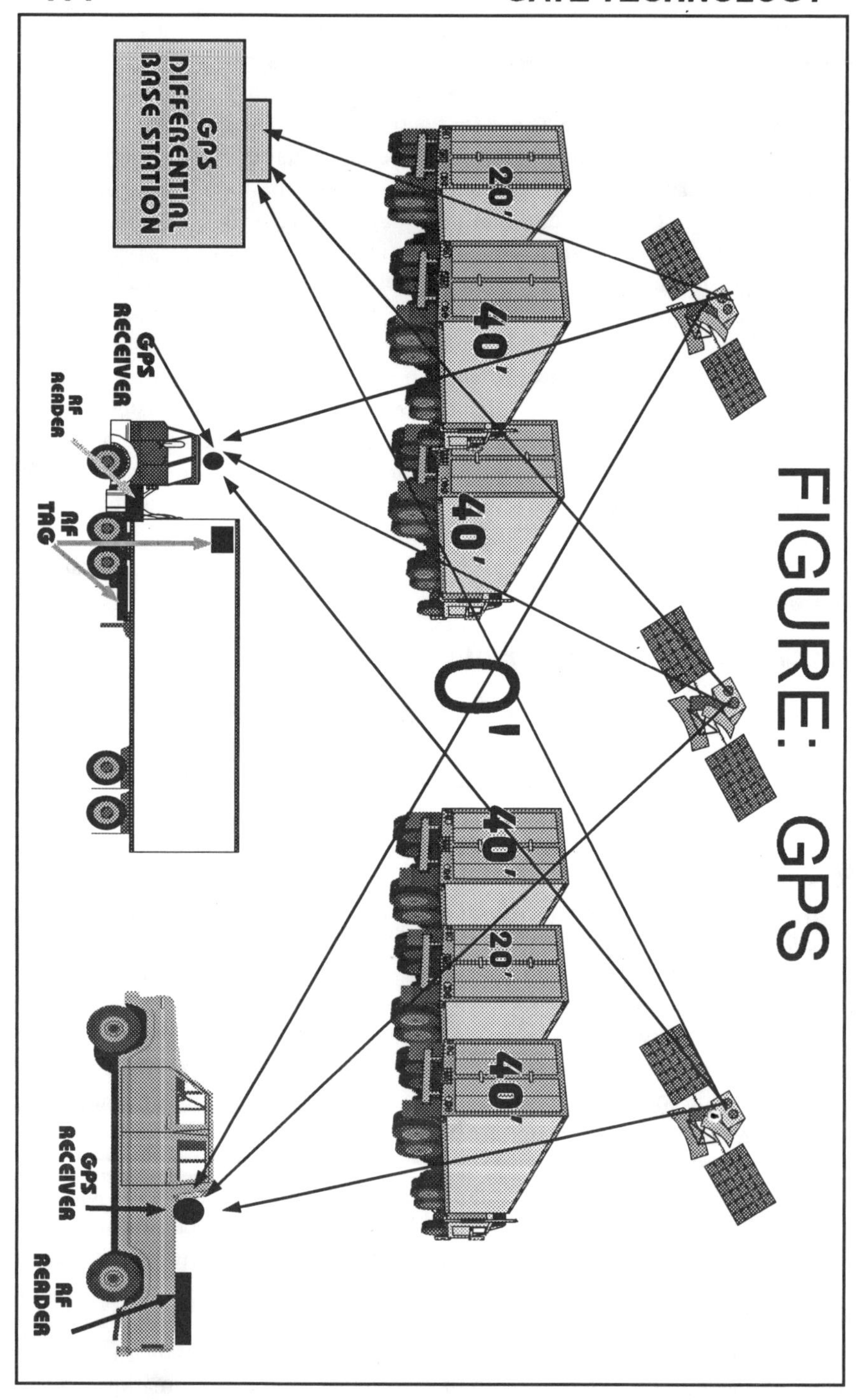

FIGURE: GPS
GPS DIFFERENTIAL BASE STATION
20'
40'
40'
0'
40'
20'
40'
GPS RECEIVER
RF READER
RF TAG
GPS RECEIVER
RF READER

captured automatically. Then information like fuel levels, engine temperatures, and the like can be sent to the system to provide operational detail. Signature capture systems can now provide the necessary documentation sign offs for processing. All of this information can be transmitted to a host, or other computer system.

The communication capabilities of computers offer significant options for your system. The use of data radios has been proven as effective. The cellular communication systems are established as well. For longer communications needs, satellite communications, and satellite and land- line combinations can provide the necessary economics for longer distance communications.

ARTIFICIAL INTELLIGENCE / NEURAL NETWORKS

When new data and more granular information is available there may be ways to assist in making every day decisions. Often we are confronted with making decisions based on subsets of the data. Even when the data is available, we often place artificial constraints on the data to make the decision easier to determine. For example, we have dispatch personnel aligned with state and city boundaries when an entire country view might be better. One of the key reasons for this is the conjecture that a given dispatcher can handle only so many transactions per time frame. The dispatcher can handle only a certain amount of trucks or pickups. Computers can handle many more transactions in a very short period of time. The computers follow a set of rules that we use. Rule based systems can produce faster more reliable suggestions for action than the human mind can keep track of in the time frame allotted. Experts tell the computers how to react, and then the best decisions of the experts can be employed in the computer system to help us make decisions. This is often referred to as artificial intelligence.

The applications possible for this discussion are many. With accurate and timely location information, planning systems can be implemented with even higher probability for success. RF/ID systems can provide the accuracy and the GPS and computer systems can get the data to the planning computer. Dynamic route planning systems are possible. Predictive fleet allocation systems can be implemented. Incident management systems and contingency planning systems will be more effective as well. The accurate data made available can provide systems that more economically schedule resources such as equipment and people.

Maintenance is one aspect but think of the systems that could be effective when a vehicle enters the gate. An eminent engine failure read into the system could demand immediate action. A system might be invoked to see how best to replace the vehicle and maintain customer service all at the same time. Of course, equipment allocation and contingency planning systems can be implemented for effective decision making.

With the proper rule based systems there are some very significant decisions

FIGURE: ARTIFICIAL INTELLIGENCE APPLICATION DIRECTIONS

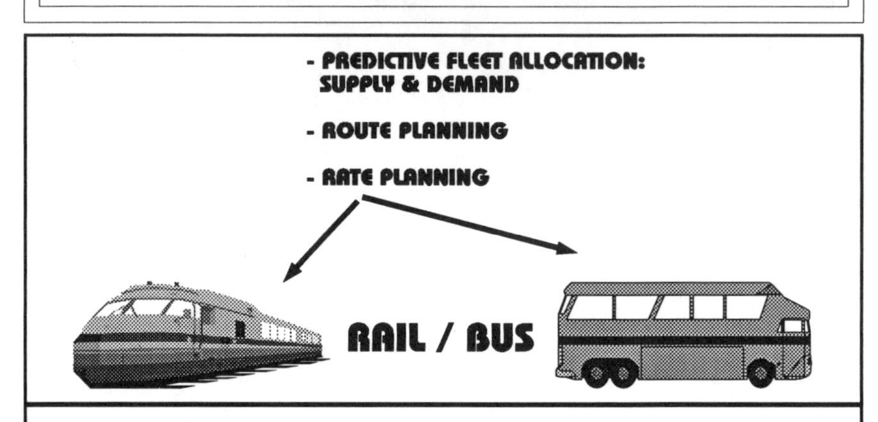

- PREDICTIVE MAINTENANCE SCHEDULING
- DYNAMIC ITINERARY
- EQUIPMENT ALLOCATION
- CONTINGENCY PLANNING

MAINTENANCE

LOCAL TRAFFIC

CONTINGENCY PLANNING
DYNAMIC RESOURCE SCHEDULING

DYNAMIC ROUTE PLANNING

INCIDENT MANAGEMENT

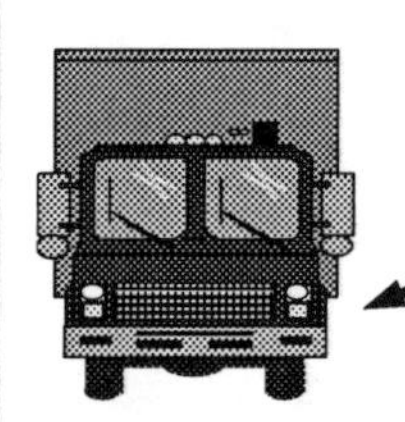

to be made, But it does not stop here. If the computer program can be given routines to analyze the situation based on HOW A human would view the problem, another artificial intelligence can be achieved. The computer routine can learn how to be more effective. In the inspection systems described above the computer could be taught to look at the size and shape of the container. 20, 40, 45, 56 foot trailers are available. Each might be smooth or ribbed, and if ribbed there may be a certain spacing between ribs and on and on. The computer could also be told how to find these characteristics. If it discovered a 65 foot trailer, it would simply add this 65 foot length to the possibilities. So the computer is said to be thinking. This function is referred to as a neural network.

Neural networks learn as the process continues. In voice recognition systems, the more the driver speaks to the system, the more patterns are available for analysis, and the easier it is to determine whose voice is being transmitted. Neural networks and artificial intelligent systems can use the accurate and timely data collection of radio frequency identification systems to assist us in the decision support processes.

MULTIMEDIA GATE SYSTEM

When you put it all together, technology at the gates and in the yard can make for revolutionary change. A gate can simply record the identification code and open, or can include much more. Cameras can record the activity around the gate. Weigh scales, voice detection and speech recognition systems can assist in the data collection, and the communication of gate transactions. The combination of these capability is called a multimedia gate.

Possibilities for the multimedia gate systems for transportation are shown in the Figure: Multimedia Gate Systems on page 108. The application, the technology and how the technology can be used to improve the process, and the benefit to the company is shown.

SUMMARY

Gate systems are everywhere! RF/ID can significantly help improve gate productivity today. There is a significant future for RF/ID as memory expands and standards are set. RF/ID helps other technologies become more effective through the use of accurate and timely identification. When RF/ID is used in cooperative combination with computers and GPS, there is significant progress to be made. More information provides for a better competitive environment. Imagine the possibilities!

FIGURE: GATE TECHNOLOGY ELEMENTS

VOICE PROMPTING
SPEECH RECOGNITION

RADIO FREQUENCY

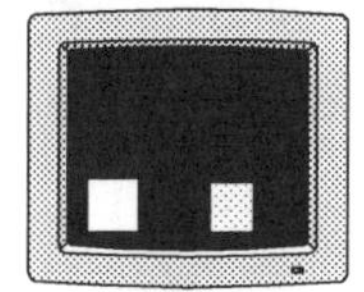

TOUCH SCREEN

VIDEO

1-2-3-4-5

OCR

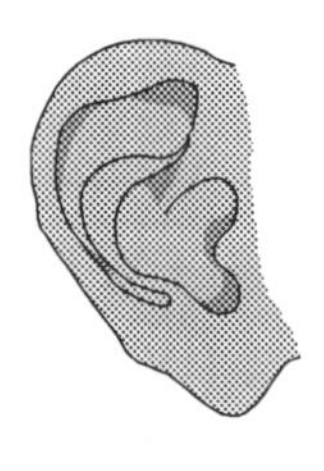

AUDIO
DETECTION

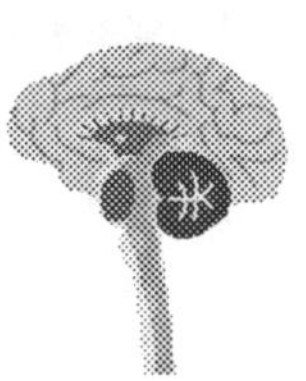

NEURAL
NET

Now,
is the
time

TEXT

OPTICAL
SCAN

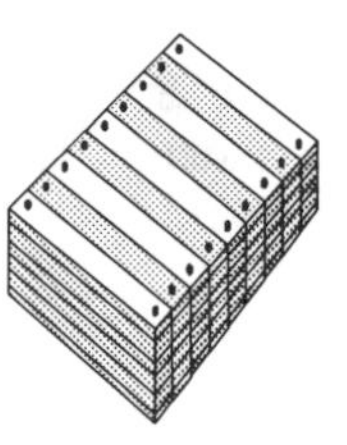

DATA

SCALE

FIGURE: MULTIMEDIA GATE SYSTEM

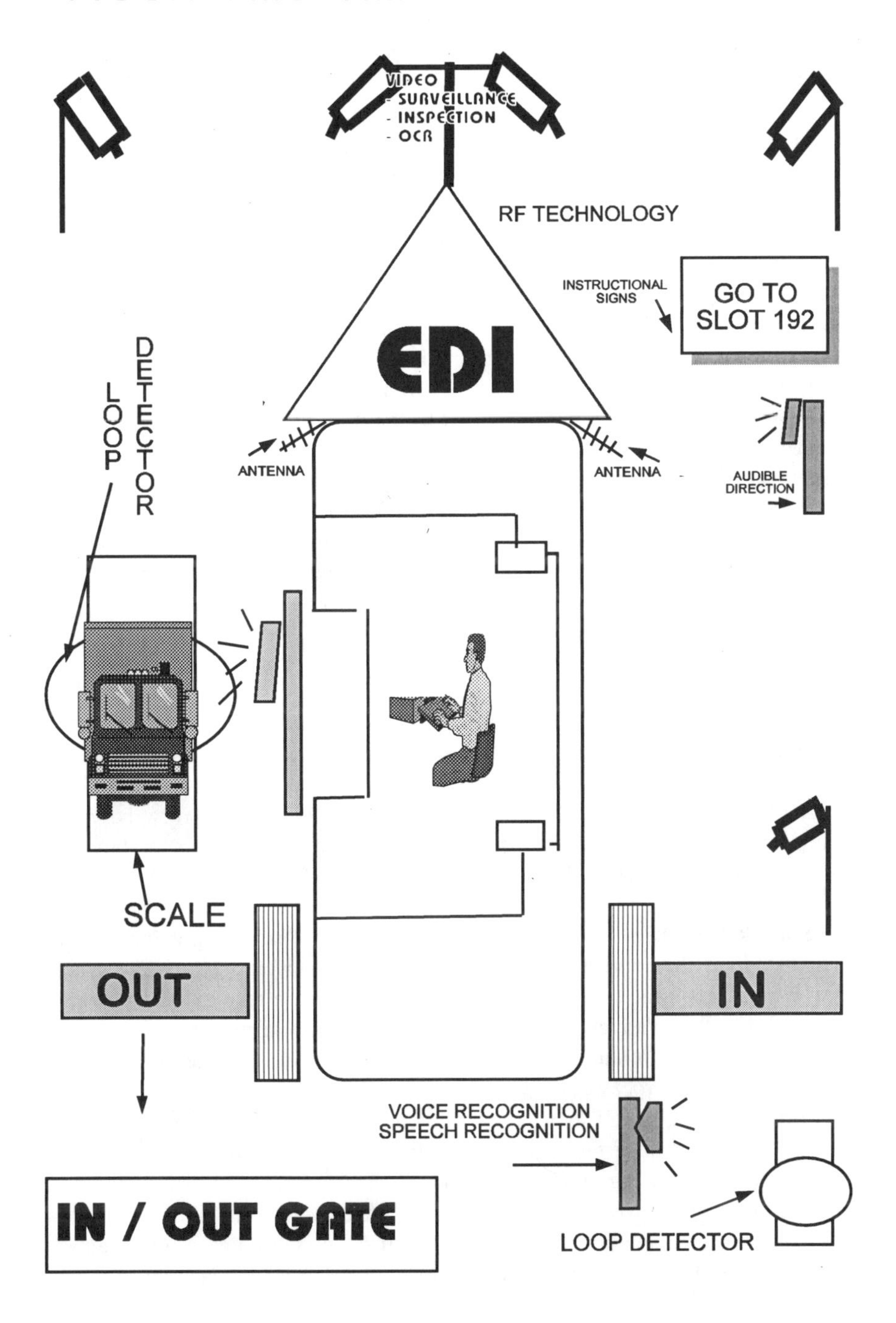

CHAPTER XI

TOLL

Toll roads and bridges have taken on a new meaning. Their importance is increasing due to the revenue, and there is the convenience to the traveler. RF/ID or AVI is a critical application to the toll market.

In the 1950's you could get on a toll road, pay a few cents, and use the relatively congestion free roadway. No wait, not much hassle and certainly, high quality roadways with smooth surfaces, and well maintained in all kinds of weather. Since that time the population has swelled, more vehicles are pounding the pavement, and the situation has changed. While most toll authorities have worked diligently to maintain and upgrade the toll infrastructure, peak period traffic resembles a parking lot at the toll collection interchanges. The concept of queue time is being addressed.

In the early days we collected fixed fees. Cash transactions in amounts easily available in denominations equal to the coin of the realm. Each driver needed to stop at a toll gate to pay. To speed the process, a ticket was dispensed at the entrance of the toll road. The ticket designated the costs for the various exits. This use charge became a way of life. It was a good way to pay for new roadway facilities. More technology was introduced as customers demanded better service and toll authorities required more accountability from the toll gate collection.

TOLL TECHNOLOGY EVOLUTION

The tickets were upgraded to provide data that could be read from ticket machines. This provided audit trails and reconciliation routines for tickets and even for some of the revenue. Displays were added to the ticket booth to inform users of the amount owed. The fee schedules then could even be more complicated. Toll collectors would make change for the drivers but had difficulty keeping up with the traffic. So more lanes were added. With more service lanes, the long lines were shortened, but traffic grew and new ways were needed to assist in the problem. If correct change only lanes were installed, then the service time for the toll collector would be much smaller than the time required to take the money and validate the amount of the change required. For the relatively minor inconvenience to the driver, that of insuring they had correct change with them, the correct change only lanes revolutionized the process.

Correct change gave way to automatic coin collection machines. Here you simply stop throw in a few coins and when the machine was properly fed it would open a gate or turn on a green light. This also improved service and had an overall cost improvement as well, but as prices went from a few coins to a few dollars a new solution was needed. Toll collection coins were invented. The coins could be purchased at a toll store, sometimes a bank, and always at a gate from a toll collector. One coin was purchased for the amount of the toll. This was a great application for toll bridges. Coin machines could be used, but with the tolls above a dollar, the correct change only lanes with manned toll collectors was again an improvement for those people who did not purchase the special coins. What we see today with all of this improvement is still in need for more improvement.

When people discuss toll roads and bridges today, many images come to mind. The blinding sea of red tail lights, the flickering yellow turn signals, slow or stopped traffic, are all images that come into mind. Some toll roads have collection stops every few miles. All of these stops and slow downs cause unwanted wear and tear on automobiles and trucks as they maneuver through the maze of lanes and collection gates. Untold gallons of fuel are spent on unproductive driving and the productivity of the drivers in their work place is significantly impacted. One wonders how much money is spent on breaks alone for the efforts needed to pass through one toll lane in the city. How much pollution would be reduced if the toll lanes are eliminated? Safety issues come to the forefront as well. All of these issues are important, safety, fuel efficiency, air quality, and productivity. Any one issue is enough to cause a change.

ETTM - ELECTRONIC TOLL AND TRAFFIC MANAGEMENT

Electronic toll systems implemented from as early as 1989 have been proven to have significant advantages in the management of roadways. If the objective is to get as many vehicles as possible through toll collection points, you might find that the easiest way of solving the problem is to simply build more gates. But there is limited real estate for gates. Also, the more gates you have the more barriers and sign posts you have. These things are contrary to improving safety. There are two key things needed in a replacement system for toll roads and bridges.

A. Fast Processing Time

B. Flexibility in Information Flow

As a solution, using electronic toll and traffic management has fewer barriers and can facilitate non-stop driving. The time to process is fast. Why? There is an electronic transfer of funds in the equation, that is why. Cash is not passed or exchanged at the toll gate with ETTM. Cash transactions might happen through the mail or the bank or even a drive up toll authority window but not at the ETTM vehicle pathway. However, this alone might not be the answer.

As with the other great inventions, this system might make an improvement only to be negated by additional traffic. With more traffic fast processing times would still provide for queue delays. A concept known as congestion pricing has been introduced. This means that the same physical facilities of the roadway are used but one price is for slow times and a higher price is charged during rush hours. The pricing structure is not limited to two tiers but is

WAIT FOR
GREEN

STOP
PAY TOLL
STOP
PAY
TOLL

STOP
PAY
TOLL
STOP

limited to the information flow of the collection system. ETTM systems have the ability to change toll prices dynamically and to collect the different charge from the system either through back office billing or transactions using the read/ write technologies of transponders. Automated vehicle identification, AVI, is a powerful tool for the ETTM marketplace and has significant non-contact, close proximity advantages over optical, magnetic, and touch technologies.

CONGESTION PRICING CONCEPT

This concept can apply to toll roads and toll bridges. In some cities there is also the concept of entire areas, or city blocks that charge a fee to vehicles entering the area. The idea is that when the demand for entry or use is high, charge higher rates. When demand is low, charge lower rates. Supply and demand will decide the use. Higher rates during peak times can cause drivers to alter driving habits and thereby balance traffic loads. This not only optimizes revenue but also increases the overall use of the traffic system.

AUTOMATED VEHICLE IDENTIFICATION (AVI)

This concept uses the radio frequency transponder placed on or inside a vehicle to automatically capture a code that represents the vehicle, or the user, for toll collection. The benefits of AVI are significant.

BENEFITS OF AVI

Increased use of the toll facility is a big advantage. 800 cars can pass through a manned toll lane. With automatic vehicle identification, the number increases to 2500. That is an increase of 312%. The trick is to get the feeder systems to accept that kind of traffic flow, but there is significant potential. There are other advantages as well. Increasing the number of vehicles through an AVI lane means less air pollution. There are estimates of up to 15% reduced emissions using this application. There are other benefits as well. There is the estimated potential of two billion gallons of fuel saved annually, and significant savings on the wear and tear of the vehicle. It is also a time saver for employers, and employees.

To collect the toll, reader systems are placed at gates or in the roadway to communicate with the transponder. Several types of transponders have been tested and used for this critical application.

A.	Read Only	TYPE I
B.	Read/Write	TYPE II
C.	Read/Write Modified	TYPE III

READ ONLY

In the Read-Only, or Type I transponder, the identification is captured as the vehicle comes close to the antenna. There is a simple routine that verifies the identification as valid. Simple table processing updates the table for valid and invalid identification codes. Valid tags get a green light. Invalid tags get a red light and an enforcement system is evoked. If there is a low balance that too can be indicated via a yellow light. Since there is significant flexibility in the system through hardware features and programming, most functions are possible.

READ/WRITE

The Read/Write system provides additional functions, not just the mere capture of the ID code. With a Type II transponder, fields such as entry and exit codes, inspection reports, balance and so on can be kept on the tag. The dynamic area can hold a balance. For pay ahead systems, the balance is available in the transponder. As the vehicle with the tag approaches the gate, or lane, the transponder is read, and the system determines whether there are sufficient funds. If there is, the system deducts the amount from the balance, and writes the new balance on the tag. When the balance is low, the system can communicate with the driver via a lane sign. If the balance is insufficient, a red light can indicate the unauthorized access and invoke an enforcement system. The Read/Write system has many possibilities available to improve data collection, and data processing required to improve the flow of traffic.

READ/WRITE MODIFIED

Consider a Read/Write tag and add the ability to insert a credit card sized device into it. In fact, the concept is that it might be a credit card or perhaps what is called a debit card. Debit cards are pay ahead cards; credit cards are pay behind or after the fact type cards. Toll authorities like to get paid ahead of time so use the pay ahead concept.

SMART CARD

The Smart Card is a device that has memory, electronics, a computer, and the Smart Card can execute a program for security reasons, or other applications as well. It can be the size of a credit card. If the users of rental cars had Smart Cards as part of their rental, they would insert the card into a device in the vehicle, and this Smart Card could identify the user, and keep track of all transactions including fuel and tolls. This could work and work well.

READ/WRITE MODIFIED SUMMARY

When the credit card, debit card or smart card is part of the system, or even if all three are used, the system takes on even more function than previously thought to be practical. Type III transponder systems have their place as well. This concept is tied to a radio frequency system to provide the communication from the transponder device in the vehicle and the reader system at roadside or lane, gate or toll booth. The system has application potential that can be significant to all involved. Customers like the security; authorities like the productivity, and everyone would enjoy the potential for cleaner air. The Type III tag can communicate with the driver through function lights and the insertion of a card.

PAYMENT CONCEPTS

Pay ahead systems are preferred over billing systems for all three tag types. There have been many options discussed for payment. The customer may acquire the tag from a bank or other financial institution. The customer would then set up a tag account and pay a certain amount to the account. at regular intervals. This could be transacted in different ways based on the tag type used.

1. Read Only: The account balance is updated and a table of amounts or sufficient fund indicators are sent to the system on a periodic basis. The drawback is that there is a significant amount of maintenance required and the timing of updates for a system that is in use 24 hours per day has little appeal.

2. Read/Write: The read/write tag can help here in a significant way. The balance can be kept on the tag. The drawback is that each customer has to go to a special location to add money to the system.

3. Read/Write Modified: This read/write modified technology can offer some advantages similar to the read/write. The balance is available from the tag. In the instance when a credit card is used in the modified connection the concept of validity and credit worthiness tend to be a drawback to the benefits.

Pay behind systems tend to be like a credit card. There are toll transactions accumulated against a credit line and a bill is sent to the user. Again all three tag types can be implemented but with different processes for the controls. All of the tag types can capture the transactions to bill later. The read/write and read/write modified can keep a maximum use or remaining balance field. The read/write modified tag can send the transaction on as if it were a credit card charge. The drawback is small credit card transaction amounts. Over the course of a month, a commuter might rack up hundreds of dollars of charges, but if they go to the credit card company at $0.50 per, the transaction processing may eat up the profits.

There may be systems that require both pay ahead and credit systems. Certainly, there are arguments for and against this concept. When the customers are grouped in categories, certain categories might have different requirements and different risk structures. The public might be on the pay ahead system, while frequent drivers, like commuters, and rental car customers might pay on credit or some combination. Drivers can add to a low balance with cash at a tollbooth or at an AVI station located in the plaza area, by personal check, or through automatic credit card billing. Alas, what about the cheater?

VEHICLE ENFORCEMENT SYSTEMS

Vehicle Enforcement Systems, (VES), have been around for a long time. The first one, when a patrol car with an enforcement officer sits and watches for violations, is still in use. However, the term VES implies more than that today. Usually there is a camera involved.

The VES consists of an interface to a controlling computer. There is a camera for each lane, and a triggering device to insure a picture is taken at the appropriate time. The camera is aimed at the part of the lane so that the license plate is in view. The picture or pictures are stored in several ways, depending on the particular approach used. For example, there may be a video tape or a series of pictures.

The lane control function time stamps the events and when a violation is detected instructs the system to note the time and date and record the captured information. There are different ways in which the VES systems keep the images. A digital camera converts the image into a digital form for storage on a disk drive. An analog/digital system uses an analog camera for each lane and a single frame grabber for conversion into digital form. An analog system keeps the images on video tape.

Timing is critical in this process. Once the end of the vehicle is detected the image must be taken and captured. While this always sounds so simple there are problems involved. The picture is for the license plate. Capturing the license plate seems simple enough, but the license plate is relatively small and the area of concern is so large. The plate may not be centered, and are at different heights depending on vehicle type. The lane is wider than most vehicles, so the license plate might appear in a wide range for the field of view. Most cameras will not take the picture with enough pixels to readily discern the small letters on the plates. Then there is the matter of the toll collection lane being out in the elements. The day or night, sun or moon, lighting conditions offer a challenge when lights of oncoming vehicles are a part of the picture. The license might also be in the back window, or have dealer graffiti surrounding it, and so on.

CLASSIFICATION

How to automatically find the end of vehicle is an issue deserving some attention. This seems simple but there are problems. Simple wheel detectors or treadle counters are not enough. The question becomes one of how many wheel sets or axles does a vehicle have. There are many varieties. A car for example might have a trailer attached. The tong of the trailer might be big and bulky or might be a thin piece of steel. Vehicles might shadow the vehicle in front as they proceed through the lane. An eighteen wheeler, dump truck or limousine might be the target vehicle, and all of these have different wheel counts.

USES

The uses of the system needs to be reviewed in view of the technology or technologies that are to be part of the system. One issue that comes up is whether there should be one technology or one vendor. Certainly the idea of implementing one like read-only has the advantage of easily working out the kinks in the system. If a standard can be agreed upon, then there are possibilities for multiple vendors or suppliers. At a minimum, competition should help, but consider different and separate radio frequency systems. At first the inclination is to say no to such an idea. After all, who would possibly want multiple tag types running through the system?

Today, there are multiple technologies at the toll collection lanes. There are toll collection personnel taking money and making change, coin machines, tickets, and gates, and lights and wheel counters and on and on. Multiple RF/ID technologies might be there one of these days as well. Certainly there are advantages to keeping the amount of clutter to a minimum. Strategically placed equipment and planned implementation with well thought out integration strategies could make the overall system more productive.

AVI STANDARDS

There are efforts to set a standard that can be agreed upon by many of the players. A standard for AVI is an infra structure question as well as a specific productivity question. There have been tests conducted by the toll authorities, and there has been implementation by toll authorities. Major companies, who manufacture the equipment, lobby hard for their version of the solution. One of the biggest issues comes from owners of fleets. They have requirements beyond the toll collection application and are concerned about having multiple transponders on the vehicles. The hazardous material enforcement authorities have particular requirements that this tag technology could help address. Border crossings and customs officials could also gain from part of a standard.

It is helpful to the process that the ATA and ISO standards for intermodal containers have been approved. While the ultimate standard for the application might be different there is a starting point. Tests are conducted by various toll agencies to help understand the technical issues to be addressed. The Inter Agency Group, IAG, in the Northeast completed significant tests of major manufacturers' products. California's CALTRANS standard has been considered for a national and even international standard.

CALTRANS AVI SPECIFICATION

The California Department of Transportation has been extremely active in the evolution of automatic vehicle identification. California was one of the earliest states to recognize the need for a non-proprietary specification and defined requirements, and signed them into law in September, 1990. Early specifications for the toll tag had significantly changed the requirements, as provided by the toll tag manufacturers, prior to the CALTRANS specification. The memory size was increased. The requirement was not only for a read capability but a write capability as well. This was one of the first specifications to require the read/write tag, and it came at a time when read-only was proven as acceptable. Of particular note in stretching the limits of the technology was the requirement for a very high data transfer rate. When state-of-the-art was 9.6 KBS, the CALTRANS specification was asking for 300 to 500 KBS. While the manufactures of toll tags lobbied for their unique specifications, a modulated backscatter was the technology chosen.

SUMMARY

The toll market has several incentives for implementing high technology. Electronic toll collection has the potential to eliminate problems handling money, and it speeds the time for collection. There is money available through the IVHS programs to assist the authorities in their AVI projects. Options are available to employ the technologies and standards are beginning to surface. Major corporations have joined in the action, and the future of automatic toll collection looks good.

CHAPTER XII

INTELLIGENT VEHICLE HIGHWAY SYSTEMS (IVHS)

Roadways are becoming more and more congested. Safety and law enforcement vehicles have difficulty getting to the scene of an accident. Safety improvements are necessary, and there must be a better way to get more through the system. RF/ID can help.

INTELLIGENT VEHICLE HIGHWAY SYSTEMS

Congestion! Congestion! Congestion! What to do about congestion? Our roadways are congested, and there is little room for more roadways. Physical space is limited where the roads are needed most. It costs too much and takes so much time to complete new roads, even when it is possible to build new ones. As the population increases our demand for more highways increases. We increase congestion without improving safety. Five million people were reported as injured and 41,000 people died in traffic accidents in 1991! A country that once boasted of having the best traffic infrastructure in the world is in need of some major efforts to overhaul the entire traffic system.

The system that once contributed to high productivity is now responsible for causing billions of dollars in losses. An estimated $100 billion in lost productivity has been quoted. Traffic accidents, many caused by congestion, resulted in losses of another $70 billion. Tons of pollutants and billions of gallons of fuel are wasted by inefficient movements and idled motors that are stuck in traffic each year.

On December 18, 1991, President Bush signed the Intermodal Surface Transportation Efficiency Act of 1991. The bill known as ISTEA is in recognition of the strategic problems to be solved. The concept is to develop a National Intermodal Transportation System that is economically efficient and environmentally sound, and that provides the foundation for the Nation to compete in the global economy to efficiently move people and goods.

GOALS

The goals for IVHS in the United States are as follows:

1. Improved Safety
2. Reduced Congestion
3. Increased and Higher Quality Mobility
4. Reduced Environmental Impact
5. Improved Energy Efficiency
6. Improved Economic Productivity
7. A Viable U. S. IVHS Industry

To reach these goals a high tech approach is envisioned. There are a number of technologies, including information processing, communications, control and

electronics. Combining these technologies to improve how we react to conditions can save lives, time and money.

IVHS is not a far distant dream. In fact, there are projects underway or in use, or that are being tested to provide first-generation solutions to some of the problems. Congress has authorized an average of $110 million per year for six years to test and develop the technologies and concepts.

TEST SYSTEMS

Systems that are under development or actual testing include a variety of vantage points.

A. Traffic Conditions
B. Traffic Incident Support
C. Automatic Tracking and Dispatch
D. Navigation Control

TRAFFIC CONDITIONS

The concept is to make available to the traveler the traffic conditions as close to real time as possible. Whether the traveler is at home, office or in route, the latest information can be available so routing decisions can be made. The trick is to collect the necessary information in an automatic and responsive fashion.

TRAFFIC INCIDENT SUPPORT

Expanding capacity through more effective handling of traffic incidents is the center of attention here. Rerouting traffic flow to avoid jams and thus eliminate or reduce further traffic incidents is also a consummate concern of the authorities. Examples include the routing of emergency vehicles, sometimes against the flow for better and faster access to the injured. To find unusual conditions, video cameras are strategically placed for control center viewing.

AUTOMATIC TRACKING AND DISPATCH

Systems are needed to improve the productivity of commercial, transit, and public safety fleets. Automatic tracking and dispatch systems dynamically reroute vehicles to accommodate changes in customer needs. For example, changing the route of a bus to a stop with many passengers, or to where the demand is high based on real time information. Another example might be to dynamically reroute a city bus around the scene of an accident to keep the flow of passengers going with the least disruption. Another example might be to direct freeway traffic well in advance of the accident to detour thereby giving traffic flow systems a chance to react to the proper signaling systems.

NAVIGATION CONTROL

At the very heart of this safety minded system is the automatic navigation of the vehicle. Warning indications to the driver are possible. Communications with the environment and the vehicle is necessary for interaction to reduce traffic incidents. Path finding on the complex end of the scale to the readily available information for tourist attractions and nearby businesses are part of the solution. Routing information on a dynamic basis is possible. Two different vehicles on the same roadway mile may be given two different directions to the same ultimate destination.

BENEFITS AND RF/ID

Gaining all of the benefits from this high tech concept of Intelligent Vehicle Highways Systems is a challenge. Many of the benefits from the various technologies that might be employed have effects on the various IVHS objectives. Radio frequency identification and the ability to automatically identify the transponders has a role in the development of IVHS systems. The following table shows how RF/ID might apply. Keep in mind that some of these ideas are concepts and have not yet been tried. As with other technologies, the real capabilities have not yet been proven or fully developed.

TABLE: IVHS BENEFITS AND RF/ID POSSIBILITIES

SAFETY OFF ROADWAY	LOCATION INFORMATION, CENTER LINE, EDGE OF ROAD INDICATION
SAFETY- TOO CLOSE	END, FRONT OF VEHICLE DETECTION
REDUCE CONGESTION IMPROVED MOBILITY	ETTM, ESPECIALLY TOLL COLLECTION, ALSO TRAFFIC STATUS, SPEED AND TIMING
ENHANCE ECONOMICS	TOLL COLLECTION, FUEL TAX
ENERGY EFFICIENCY	FUEL FROM LESS STOPS
ENVIRONMENTAL QUALITY	LESS IDLE TIME
VIABLE IVHS INDUSTRY	TAG MANUFACTURERS, CONSULTANTS, INTERROGATORS

There are significant contributions to be made by the radio frequency identification technology to the IVHS program. The benefits come as RF/ID plays its role as part of the IVHS partnership team. A reliable identification with a hands free, non-contact approach is required to gain the benefits, and RF/ID can supply this function. Once the identification system is in place, there are additional uses and enhanced functions that can be added for a rich set of application possibilities. Memory in the transponders, for example, can provide a place to keep inspection records, dynamic travel information, results from weigh scales and on and on. Low cost identification makes it possible to then dynamically provide public transportation routing information, and provide the traveler with real time updates. However, given that the partnership of RF/ID and IVHS is a strong one, we must keep in mind that RF/ID is relatively new as a technology and most systems are in the first or near first generation of packaging and design. This means that there are even more possibilities for cost benefits and functional enhancements. The IVHS program has a way to address the expansion and development of the technology in the partnership approach being taken.

IVHS embodies a wide array of technologies and challenges. Soliciting public and private sectors as well as academic sources provides for fast development of solutions to the problems. The figure below shows the concept with IVHS leading the effort to coordinate the three groups.

FIGURE: IVHS PARTNERSHIP CONCEPT

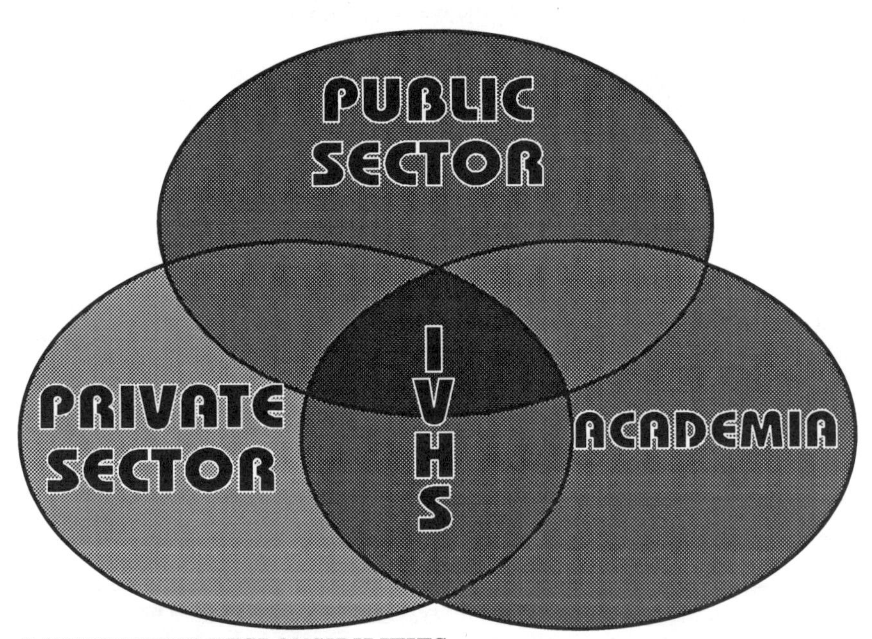

PARTNERSHIP RESPONSIBILITIES

There is a significant opportunity to advance the functional richness of RF/ID systems with the IVHS program. The responsibilities of the players from the viewpoint of the IVHS program provides the flexibility and some fundamentally new approaches that make this effort truly unique.

TABLE: IVHS RESPONSIBILITIES

IVHS AMERICA

* Maintain Strategic Plan and Clearinghouse
* Provide forum for solving problems
* Evaluate and advise

FEDERAL GOVERNMENT

* Fund and manage research and testing
* Remove institutional barriers

STATE AND LOCAL GOVERNMENT

* Deploy IVHS
* Operate and maintain IVHS

PROFESSIONAL SOCIETIES

* Develop standards
* Disseminate information

PRIVATE SECTOR

* Develop products
* Market products
* Act as research partner

FEDERAL LABS

* Transfer technology
* Provide development expertise

ACADEMIA

* Educate IVHS professionals
* Act as research partner

So, here is the framework in which IVHS America wants us to operate. The power of this effort comes from two key elements. The private sector can play a role in the research, and the federal government can remove some of the institutional barriers to accomplish this. It remains to be seen just how effective this will be. In early efforts there has been significant barriers to have the private sector join with a local government and use federal funds to build prototypes to solve a problem. Concepts like 'single source' get into the way. Who owns and has the rights to the intellectual property when private industry spends money in a sharing environment with federal funds? But, the concept is very powerful when and if the bottlenecks are removed.

Our highway agencies have been primarily oriented toward the civil engineering disciplines. In fact, with emphasis on civil engineering, one wonders if there might be very strategic funding wars in the making. Our roadways and bridges need serious overhauling if the current system is to be maintained. The money needed to support the reconstruction, repair, and the annual maintenance required to support the bridges, is a very significant part of the budget. Pressures to reduce the deficit, maintain space programs, and more military support may cause the high tech IVHS concepts to be delayed or even canceled. As a minimum, it is expected that even with high congestion the roadways and bridges must be functioning and in working order. The bridge must be standing to cross over it. All the high tech in the world will not help the broken bridge.

The high tech of IVHS might be even more important in the environment of reduced roadways, even when there are temporary closings. While there will be pressures to redirect funds, there is a genuine need to increase the efficiency of the highways and bridges currently in service. This need will drive the funding issues benefiting IVHS programs.

IVHS PROJECTS

Research and development as well as operational tests are the IVHS strategic plan. Radio frequency identification plays a major role in these development projects. Identification is an important aspect of any system, and RF/ID has some significant advantages. Real time collection systems and easy access to the data are also important in the overall use of the RF/ID technology. There are five major systems areas of concern.

1. **ATMS** — **Advanced Traffic Management Systems**
2. **ATIS** — **Advanced Traveler Information Systems**
3. **AVCS** — **Advanced Vehicle Control Systems**
4. **CVO** — **Commercial Vehicle Operations**
5. **APTS** — **Advanced Public Transportation Systems**

In these systems there will be a need for multiple technologies to feed decision support systems to effect the desired change. Hybrid systems will use multiple inputs to determine levels of examination of data required for corrective or supporting actions.

The table on the following page shows some of the ways in which RF/ID is used within the various system areas.

TABLE: RF/ID & IVHS PROJECTS

IVHS DESCRIPTION	RF/ID APPLICATION
ATMS	
-Traffic Monitoring	- ID, Speed
- Traffic Control	- Lane Preference - Ramp Control - Flow
- Traffic Management Center	- Count Statistics
- Multi Source traffic Data	- Flow, Speed
- Dynamic Routing Strategies	- ID
- Area Wide Traffic Mgmt.	- Flow
- Multi Mode Integration	- ID, People ID
ATIS	
- Navigation Software	- Periodic Updates
- Communication Alternative	- Fixed Data
- Dynamic Route, Optimal	- Near Destination Msg.
- In- Vehicle Signing	- Near Destination Msg.
- AVI, and AVL	- AVI, and AVL
AVCS	
- Collision Warning	- End Of Vehicle Signal - Front Of Vehicle Signal
- Automated Highway	- Lane & Edge Of Road
- Automated Freeway Lane	- Lane Indicator, ID
CVO	
- Weigh- In- Motion	- ID & Storage of Weight
- Electronic Toll Collection	- ID, Balance, Gate
- Electronic Record Keeping	- ID, Driver ID, Miles, Fuel, Time
- Driver Warning Systems	- Lane, Destination
- Electronic Credentials	- Driver ID, Shipper, Consignee, Bill of Lading
- Electronic Permitting	- Record With Equipment
- Hasmat Cargo	- ID, Description, Safety Officials
- Automated Inspections	- ID, Record
APTS	
- HOV Verification	- ID
- Electronic Fare Collection	- ID, Fare Balance, Collection
- HOV Guide Controls	- Lane, Vehicle ID
- Smart Cards	- Communications
- Fleet Management Systems	- ID, Miles Driven, Fuel
- Maintenance Tracking	- Status, Actions Taken
- Automated Transit Vehicle	- Lane Indicator, ID

The application of radio frequency identification transponders, and the reader equipment has some significant potential in the IVHS program. These projects stem from research projects identified and operational tests that are being conducted. The use of radio frequency identification has been tried in the toll collection application. You may desire to read the chapter on Toll to familiarize yourself with the detail concepts involved. Besides the toll collection, the notion of congestion pricing is one that fits well with the IVHS objectives.

CONGESTION PRICING

Congestion pricing involves charging the user a fee or toll based on the use of a roadway or section of road or set of roads contained in an area. The charge is based on high and low usage. For example, if the use is during rush hour it will cost more, but the use of the roadway in off peak times would cost less. Government authorities have tried voluntary work shift scheduling to assist in this effort with much success. However, as the volume increases, there is very little help from a volunteer system. For example, one company staggers work shifts to alleviate traffic flow problems, but other companies in the area do not. The chances are that the congestion build up will overlap or remain throughout the staggered shifts, so everyone is unhappy.

The concept is good. The application of congestion pricing can provide some significant advantages. Imagine if the concept were to be applied to major activities like concerts and sporting events. Arrive early and pay $2.00 to park, arrive within prime time and pay $10.00. The same might apply to toll gates that service the theaters, stadiums and auditoriums. Below is a figure that describes the concept of congestion pricing.

FIGURE: CONGESTION PRICING

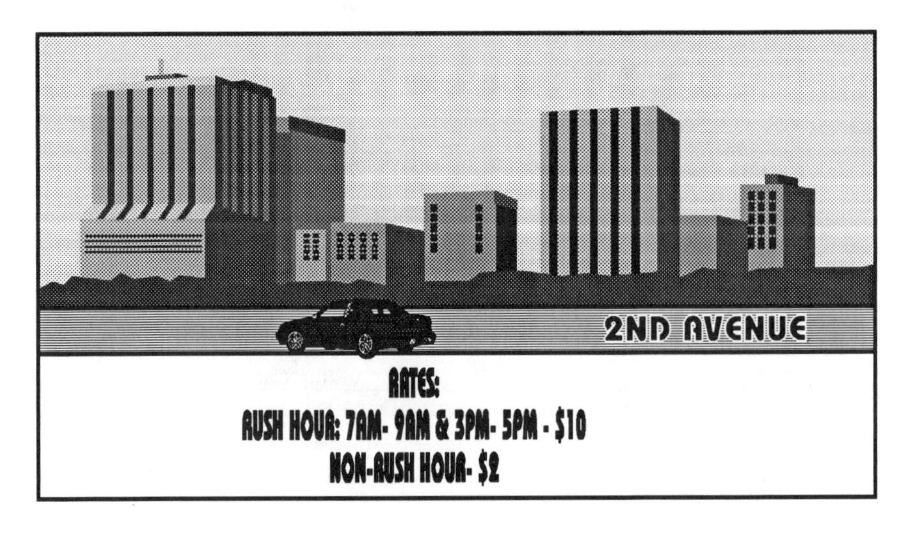

So if there is a driver on second street between 7AM and 9AM or 3PM and 5PM the charge is $10.00. However if you are on Second street between 5AM and 6AM or 6PM and 7PM the cost is $2.00. So congestion pricing is the concept here. Pay for high demand use.

OTHER CONCEPTS

Advanced Traffic Management System concepts have some particular synergy for RF/ID. The systems needed to assist the traffic flow are beneficial when integrated with other technologies. The Figure: ATMS, Advanced Traffic Management System has several concepts that should be highlighted. Emergency dispatch, ramp management, traffic management, and weigh scales are shown.

EMERGENCY DISPATCH

As shown in the Figure: ATMS, there is a plethora of technology that can be applied to the roadside. When employed, special features might be made available to the emergency vehicles for faster deployment of the servicing vehicles. When there is an accident on the roadway, an ambulance and police are dispatched to the scene. Today, they are dispatched and often they are found blocked in the flow of traffic caused by the accident. It is difficult to weave through the traffic, and it is a very, very slow process. While the scenarios are certainly subject to many combinations, someday in the future an emergency dispatch might be handled like this.

1. Traffic sensors in the roadway provide input to neural networks and decision support systems that indicate a reason for alarm.
2. Instantly, secondary high sensing retrieval of information comes into a center to help provide data to analyze the situation. Perhaps a video camera scans the area and the pattern recognition system recognizes an accident of a level four type.
3. The dispatch system receives the message from the analysis, and based on sensing data, two ambulances are dispatched rather than one or ten.
4. Within seconds downstream traffic has been diverted to alternative routes.
5. The direct route to the accident has been totally blocked with the immediate traffic following the vehicles involved in the accident. The emergency vehicles are told to enter the wrong way entrance and backtrack to the scene.
6. Phone calls to 911 begin to confirm the accident details. Help is already on its way.
7. Help arrives and lives are saved. The traffic jam is cleared with the least disruption to the travelers.

Perhaps, there are other scenarios that perhaps would more dramatically reflect the assistance that can be provided to the emergency dispatch centers. This one is given to indicate there are opportunities for improvement over today's environment.

RAMP MANAGEMENT

The concept of Ramp Management is one that has a couple of objectives.

1. Selective Preferential Treatment
   - Emergency Vehicles
   - Police
   - Car Pools
   - Public Transit Vehicles
2. Traffic Flow Improvements

The preferential treatment concept using a lane of an access ramp for special vehicles is relatively new. They could be police cars, or car pools. With a special permit, the vehicle would be given a radio frequency transponder. As the vehicle with the transponder entered the special lane of the ramp, a sensor would tell the traffic light controller to turn the light to green for this lane, and red for the other lane, given that all other safety factors were in agreement. The lane without transponders validated for the fast lane would be stopped. Traffic control officials believe this concept would encourage car pooling and increase ridership of the public transit busses.

Authorized emergency vehicles could be given the green light while the remaining traffic would be delayed. This concept does not have to be binary in design. In fact, a multiple level approach of authorization can be used. Law enforcement and safety vehicles could be given highest priority or Priority 1. This priority may be set by the enforcement officials in the course of operation. Busses and other public transit vehicles might have priority levels of two and three. Car pools might have priority levels of three and four. All other vehicles might be given the lowest priority status. Another concept is that the priorities might be effected in different zones.

A zone might be a set of ramps, or be classified by time of day. If we expand our thinking a little, we might apply this ramp concept to key intersections as well.

TRAFFIC MANAGEMENT

Sensors in the roadways and along the road sides can provide significant traffic management data. Systems like neural networks and other decision support software can assist in collecting the data, and information. Decisions like changing speed limits when the road is wet or when snow covered are possible. Assistance in analyzing the condition of the accident or the condition of rush hour traffic is often harder to discern. Using multiple inputs on traffic flow volumes and speeds can assist the operations center personnel in determining the conditions as changes occur.

Multiple sensors and input types will most likely be implemented to assist in this traffic management problem. A certain amount of redundancy is required to insure accurate inputs for the harsh environment of the roadside. Loops, RF/ID and other presence detectors can be employed to assist traffic managers in the monitoring of the roadways. Input such as the number of vehicles, their speed, and time of day will help in the decision making processes. Following particular ID's (transponders) through the system can provide specific commuter information for general message indication to other travelers. Video cameras can be employed on a periodic basis for surveillance, and can be specifically controlled when the situation calls for it. Pattern recognition techniques can be used to help control room operators look for unusual conditions for further action.

Advanced traffic management systems will monitor the roadways, determine unusual situations, and react to problems on a real time dynamic basis. For example there have been directional lanes for some time. Traffic flows north in the center lane in the morning and south in the evening. This concept assists the rush hour traffic situations. Usually the traffic signs have been permanently fixed designating hours for the directions. There have been some installations of signs that are more dynamic than permanent signs. The IVHS concept will provide for such changes on a dynamic basis and will be more pervasive. As data on the increased traffic congestion is captured, rerouting actions and directional lane changes might be temporarily employed to help keep the traffic flowing.

The concept of a vehicle management system and the idea of multiple inputs is shown on the Figure: Traffic Operations Center and Network. The use of artificial intelligence and data base inputs from multiple computers with a sophisticated communications network is highlighted. The idea of a command center is illustrated.

WEIGH SCALE

The weigh scale application is one that has needed updating for many years. Weigh scale operations invoke images of long lines of tractor trailers waiting their turn along a highway weigh scale, tractor trailers slowing down to enter the facility; and tractor trailers trying desperately to merge back into traffic. Another image is the totally vacant weigh scale with the sign "closed." This is not necessarily a good way to alleviate the crowded conditions at the scale. There are safety issues and regulatory compliance issues at stake here.

To assist in this process of weighing and to eliminate the long lines and wasted time in the queue a new concept has been developed. Have each tractor trailer fitted with an identification tag that is a read/write transponder. Certified weigh scales write the weight on the transponder, and as the vehicle approaches a down stream weigh scale the transponder is read by a roadside

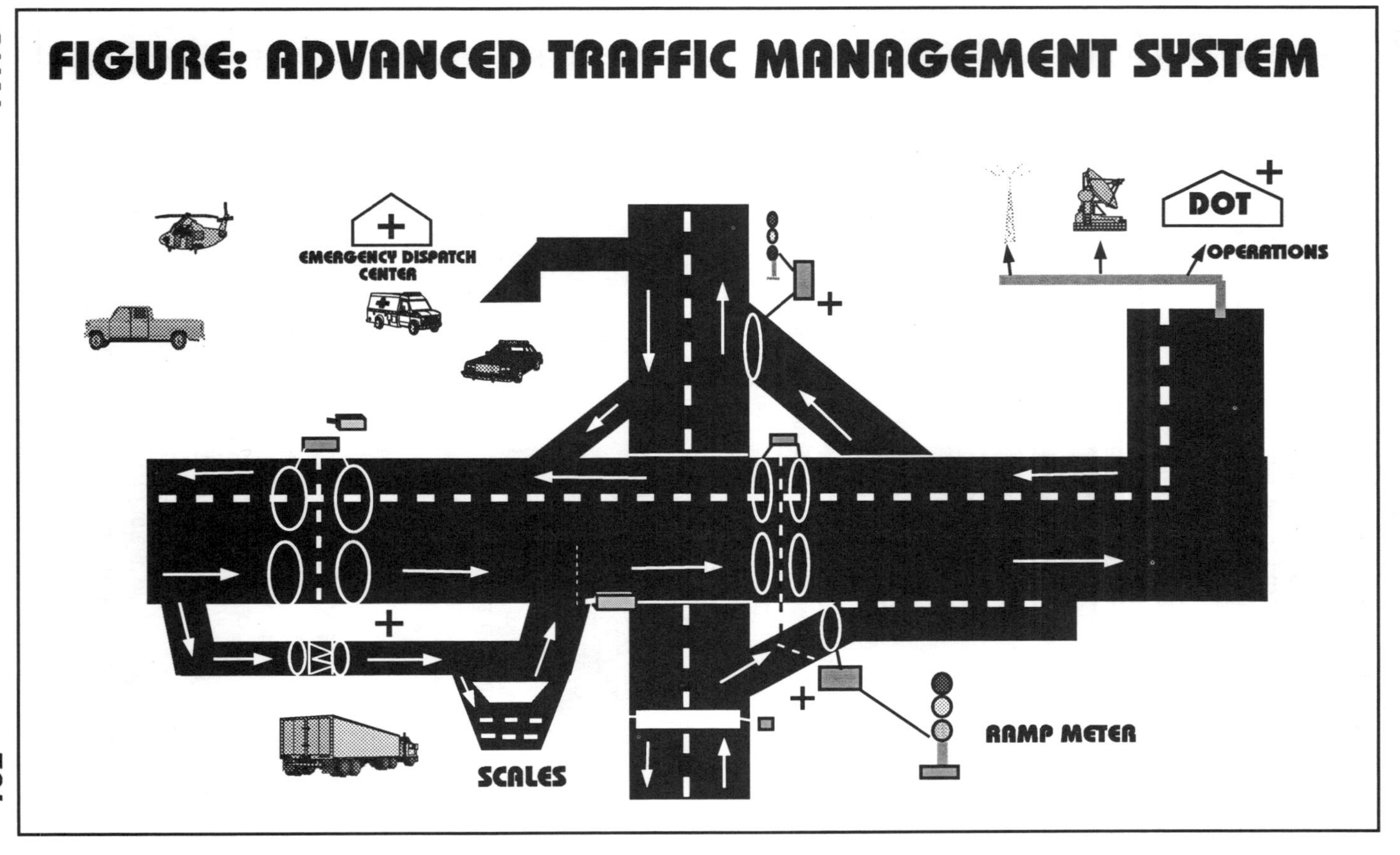

FIGURE: ADVANCED TRAFFIC MANAGEMENT SYSTEM

reader. If the time of travel and other information is in line with regulations, the vehicle is given a green light to keep moving through the main traffic lanes. If a weight has not been taken or some other information is needed for inspection, then the truck is directed into the weigh- in- motion scales, and the weight and station information is written on the transponder. If this is all that is needed, the driver is told to keep moving. If however, additional inspection is needed, the driver will be asked to stop. From time to time there will be additional "spot " inspections. This process should save significant time and effort for the drivers. Adjoining states could agree to share the information and eliminate multiple weigh station stops for drivers.

OTHER APPLICATIONS

It is possible with the RF/ID technology to supply fuel information at state boundaries. Fueling stations could write the fueling actions on the transponder. Using the dynamic tag concept, fuel levels could be provided at border crossings. With a combination of driver ID tags, and vehicle transponders or the electronic license plate, it is possible to record the driver log information on an automatic basis.

Maintenance records could also be written onto the read/write tag. This process could assist in automatic inspections and further eliminate delays for compliant operators. Authorized inspection stations could write the outcome of the inspection on the transponder with the date and time stamp information to be picked up later at a read station.

SPECIFIC IVHS PROJECTS

Following are strategic projects identified that have radio frequency identification potential. Some of these are more advanced than others but they are of significance to the IVHS program. These are primarily operations tests to be performed.

1. Test information features in rental car fleets. Use AVL and AVI.
2. Demonstrate various technologies for integrated parking and transit information and for toll debiting.
3. Demonstrate the use of electronic fare collection employing smart cards.
4. Test alternative signal preemption.
5. Demonstrate alternative technologies for automated vehicle location and passenger counting.
6. Expand test to include the integration of AVL with system wide computer aided transit dispatching.
7. In one or more truck corridors, demonstrate various AVI technologies as they apply to special needs of commercial vehicles.
8. Demonstrate various approaches to hazardous cargo tracking using a combination of AVI and AVL techniques.

9. Test a variety of automatic toll collection and automated vehicle identification techniques in a large urban area.
10. Test various ways to automatically classify vehicles.
11. Test approaches to automatically record toll and state boundary violators which do not have AVI transponders.

SUMMARY

These projects touch on some of the ways in which AVI systems are thought to be used in the IVHS projects. The future will undoubtedly uncover addition ways for the RF/ID systems to assist in meeting the objectives of the IVHS program. The RF/ID technology appears to have a very significant future in the IVHS applications.

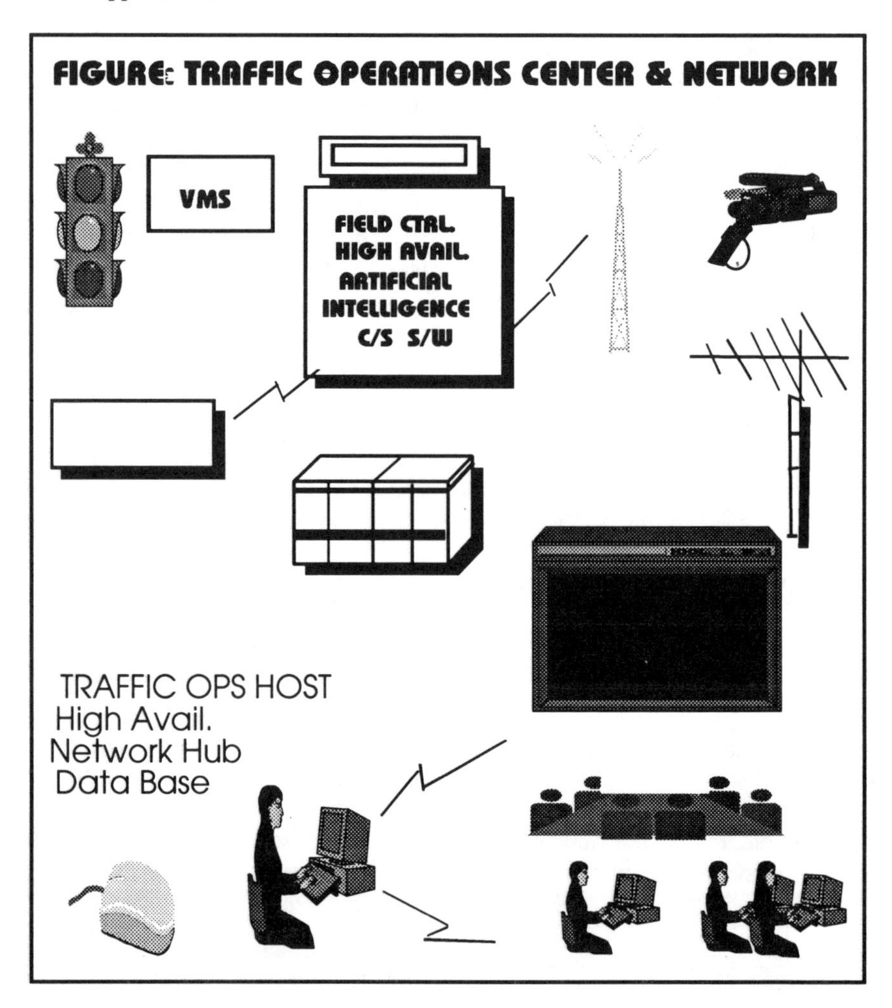

CHAPTER XIII

AIRLINES, AIR CARGO AND AIRPORTS

An interest in security, a fluctuating volume of traffic, and limited available facilities provide for a significant opportunity for radio frequency identification. Airlines and airports fit this category.

Is radio frequency identification really ready for Airports and Airlines? RF/ID could have a big effect on the identification of people, places and things, but it has been slow going. Japan Airlines tried to identify unit load devices, LAX put in a limo and taxi tracking system, and just about every congested area of the airport has been looked at for more efficient ways to alleviate the congestion and track important events. Once in a while there is even an attempt to think of customer service options, so there is plenty of opportunity for applications, but before you examine them, you must look at the standards and the technology.

The International Air Transport Association has recommended practice 1640, Use of Radio Frequency Technology for Automatic Identification of Unit Load Devices. This was a very significant step toward some acceptance of the technology. With so many potential areas of concern in an airport or airline operation, certainly there would be many more practices to follow! Are people worried about the interference with the voice and data communications frequencies used to control the flights?

Lets look at the electromagnetic spectrum for the area of concern. The Federal Communications Commission and the government agencies of all countries are involved when you start depleting these frequencies. On the chart, on page 138, that shows the electromagnetic frequency spectrum you can see how the spectrum has been allocated. The symbol * on the left is used to point out where some of the primary areas of concern are. The point is that there are frequency ranges allocated for air navigation and communications. Radar, cellular and RF/ID parts of the spectrum are close but allocated nonetheless.

So what is the particular problem or problems? One is that the various countries have different parts of the spectrum for use by identification systems. The U. S. has 915 -928 MHz, while Europe uses 2450 MHz and others use 512 MHz. Fortunately, there are radio frequency transponders that can operate in multiple frequency ranges.

Once these frequency issues are resolved, consider what other primary characteristic is needed. For the most part, the radio frequency identification systems have used a choke point advantage for reading the ID tags. A gate or a lane or some way of channeling the thing to be identified was required. This was an appropriate advantage for security applications and for handling traffic lanes, parking lots and automating taxi service. However, along came the notion of reading all transponders in the field of view. Using a Time Division Multiple Access protocol, TDMA, all transponders could be read.

PHOTOS BY GERDEMAN

RADIO WAVES

EXTREMELY HIGH **300,000 MHZ** **- Future Use** **- Short Range Military**	**SUPER HIGH** **30,000 MHZ** **- Commercial Satellites** **- Microwave Relay**
ULTRA HIGH **3,000 MHZ** ***- RF/ID** **- UHF Television** **- Microwave Ovens** ***- Radar** **- Weather Satellites**	**VERY HIGH** **300 MHZ** **- FM Radio** **- VHF Television** ***- Police & Taxi** ***- Air Navigation** ***- Military Satellites**
HIGH **30 MHZ** **- International Short Wave** ***- Long Range Military** **- Ham Radio** **- CB**	**MEDIUM** **3 MHZ** **- AM Radio** ***- Air & Marine Communications** **- SOS Signals** **- Ham Radio**
LOW **0.3 MHZ** ***- Air & Marine Navigation**	**VERY LOW** **0.03 MHZ**

CHART: ELECTROMAGNETIC SPECTRUM
(Source, IVHS America)

Look at a typical gate the next time you fly. There are all kind of devices and equipment in the vicinity of the aircraft prior to departure. There are cleaning crew trucks, food service vehicles, fueling devices, and carts with and without ULD's. It would be nice to know the time and length of stay for these pieces of equipment. My count looking out the window of the terminal was 47 pieces of equipment to service the aircraft before boarding. As I was gaining my 1 million miles of air travel I noticed a range of 17 to 47 pieces. Of course, the type of airplane is a factor for the equipment needed. Air freight terminals can have 10,000 pieces at an air hub, all moving about in a five or six hour period. Airports and Airlines have similar numbers. There are problems when the weather is bad or there are shortages of equipment during peak demand, and it would be nice to know the location of the equipment. Tracking the crew and service personnel who serviced the aircraft is also important. The wide-area protocol could help in this situation.

Antennas, strategically placed, could determine the equipment located in the area of concern. One such area could be the gate. Data collection devices could log the equipment passing by or stopping near the aircraft and log the data for review. Using neural network software routines, and other tools, and immediate departures from normal processing, might help security officials control the processes. Operational management could be in a better position to make informed decisions.

Fundamentally there is a need for accurate and timely data to feed the systems that will allow a higher level of customer service, increase productivity of the operations, and increase the utilization of the facilities.

RF/ID AIRPORT/AIRLINE PROCESSES

Let's look at how RF/ID might be applied in the airport environment. The concept for use might have RF/ID systems employed in the choke point process or using the TDMA protocol for area surveillance. Both are significant and useful.

*Crew's and other employees could use the low frequency ID card for hands free identification.

*An arriving passenger might receive better service from the taxi driver that has been automatically dispatched from a queue formed from the vehicle ID tag. High frequency could be used here.

*Commuters and frequent flyers might gain access to special parking lots. Others might be directed to the parking area that is open and available.

*Parking garages might automatically collect the tolls based on AVI systems.

*Baggage might be sorted, handled, and tracked.

*Once in the terminal, airline gate agents and attendants could immediately know that the passenger is in the building or at baggage check in.

*Shuttle flyers might use RF/ID tickets to board without any additional paperwork other than the billing statement.

*Passenger displays could inform travelers automatically of gate and time information as they pass special kiosks.

*Baggage and ULD identification could be automatically recorded.

*When a passenger deplanes, RF/ID could help retrieve the baggage immediately.

*At intermediate stops and destinations, the baggage could be sorted using RF/ID.

*During rushed and stressed times, like heavy rains, snow and fog, equipment could be located and a readiness status more easily obtained. Ramp control is possible.

*Air side security gates could monitor the passage of vehicles, such as service trucks and fuel trucks or security patrol cars and police.

There are many more applications that might fit. RF/ID is really at the beginning of it's life cycle. Imagine the possibilities!

AIRPLANE CAPACITY

Airplanes seem to keep getting bigger in terms of the passengers they carry. With a slight increase in size, an airplane can hold additional passengers. As the mechanical and other technologies improve, the planes can be operated in a

PHOTOS BY BURNAM

very flexible way. They can maneuver in close quarters, and so, do not always put additional stress on the layout of the airports they will serve. The following table shows the passenger counts of some aircraft.

AIRCRAFT	NUMBER OF PASSENGERS
DOUGLAS DC9	**98**
BOEING B737 - 200	**107**
BOEING B737 - 300	**128**
DOUGLAS MD88	**142**
BOEING 727 - 200	**148**
BOEING 757 - 200	**187**
BOEING 767 - 200	**204**
BOEING 767 - ER	**218**
LOCKHEED L - 10H - 500	**241**
BOEING 767 - 300	**254**
LOCKHEED L - 1011 - 250	**269**
LOCKHEED L - 1011 - 1	**302**

(Source system route map Delta Airlines)

TABLE: AIRCRAFT PASSENGER SIZE

AIRPORTS

Airports are divided into different areas of importance. The figures on pages 142 & 143 give us a graphic view of the airport. One notable category is the public or land side designation. The land side includes the roadways that access the airport, the parking facilities and the ticket and gate areas where the passengers and visitors can go. Contrast this area to that of the more secure area of air side.

LAND SIDE

There will be significant increases in airport traffic through the 1990's into the 2000's. Increased traffic will cause congestion and other security problems that need to be addressed by the airport authorities. As both business and general public travel increases, additional stress is placed on traffic patterns and the use of airport facilities. Peak periods such as holidays have already shown the enormous shortages of facilities that exist. The airlines can take on new and larger aircraft to handle the increases, but the airport authorities are faced with other problems. One specific problem is the shortage of land available for expansion to meet future needs. Expansion is not as likely as the optimization of the current facilities. The alternative, building new airports at other locations, will sometimes occur but it is highly unlikely that all of the current airports will close down.

Supply and demand for parking space continues to be a factor. Parking costs go up and down, and certainly this is one lever the airport operators have. Raise the price and force more off-site parking for the airports or increase taxi flow. However, taxis are in need of help as well.

PUBLIC PARKING

Information displays assist local travelers to find the available spaces in the parking areas. These systems are available. "Lot Full" signs to "Level 3 Open" is the kind of detail needed. Customer service is a key problem. Publicly elected officials need to satisfy their constituents. The concept is that a special AVI application system might at least make the trip more convenient when there is parking.

The airport parking authority could use the RF/ID system to open and close gates, both in and out gates. Date and time information combined with the identification could provide input for automatic billing against credit or debit cards, or statement billing. This could speed up the process to pay for the parking and provide for significant automatically collected operational data for the airport parking authorities. When lots are full, signs could be automatically updated. Volume statistics could be saved for further analysis. Automatic pay lanes could speed the flow on return trips, but special service lots might also need to be employed.

SPECIAL SERVICE LOTS

As a solution, special frequent user identification systems might reserve space for special fees. Airlines might consider frequent flyer parking to inspire loyalty. Here again supply and demand factors enter into the picture. Corporations and individuals might pay premiums for parking, if there is a time savings and location benefit from the process. Imagine having the AVI system open a gate, and find a parking spot a few feet from the airline terminal! This might be more than what we can hope for, but it is a possibility. Special ID tags could be used for these facilities. In fact, a card similar to the one used for public parking could be used and even be interchangeable with the concept of an "airport card", So when one lot is full, signs direct you to another. Frequent flyer ID's could be checked against passenger lists and airlines notified of the passenger's arrival.

ARRIVALS

When a plane arrives, there is another influence on the land side traffic. One of the first things a passenger needs is transportation from the airport to his or her destination. This involves transporting the passengers away from the airport, typically including: Taxicabs - Hotel Limousines - Rental Car Busses

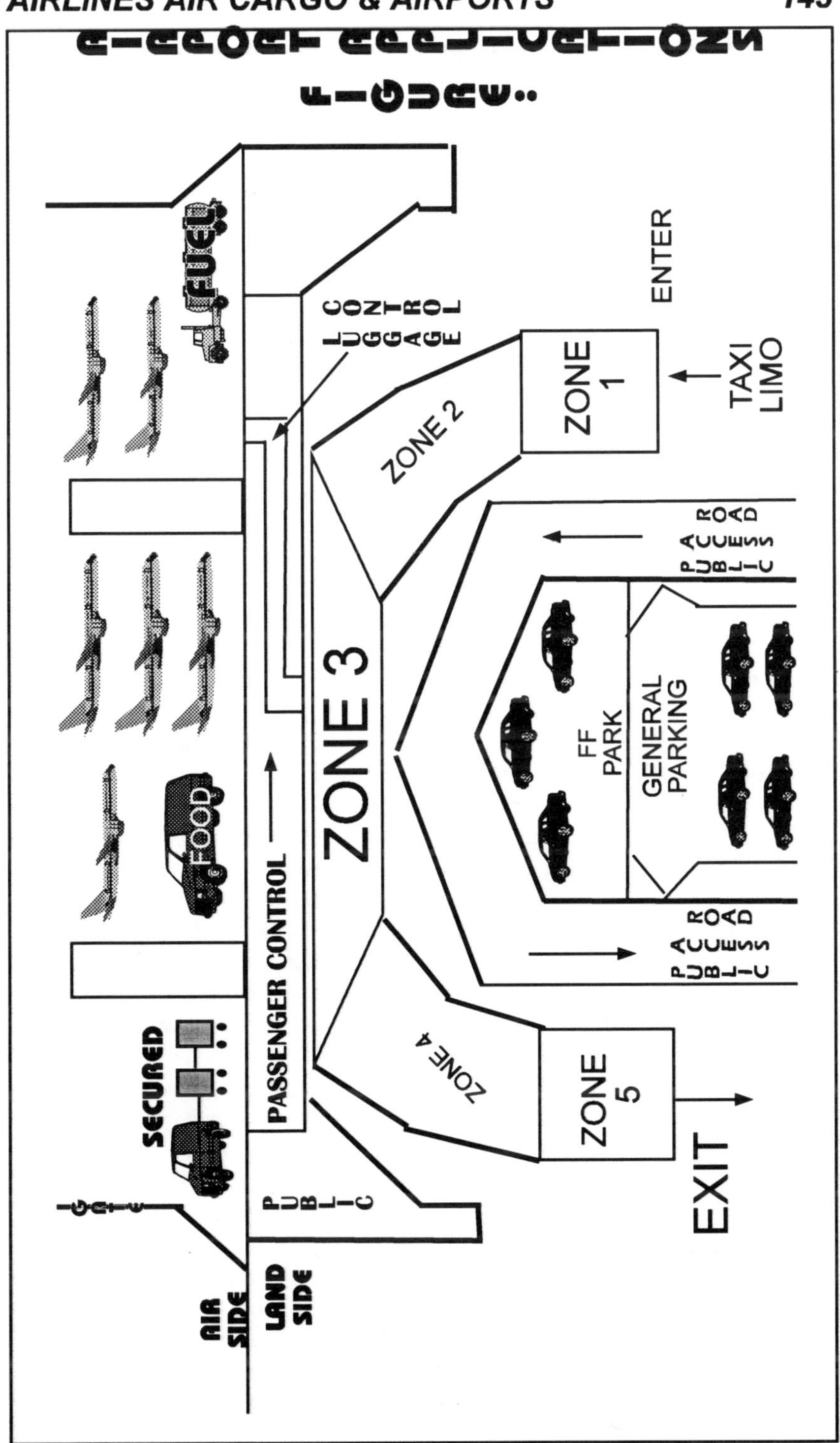
FUEL
LUGGAGE CONTROL
ENTER
ZONE 1
TAXI LIMO
ZONE 2
PUBLIC ACCESS ROAD
ZONE 3
FF PARK
GENERAL PARKING
FOOD
PASSENGER CONTROL
PUBLIC ACCESS ROAD
SECURED
ZONE 4
ZONE 5
EXIT
GATE
PUBLIC
AIR SIDE
LAND SIDE

The problem is that as traffic grows, congestion increases to the point of gridlock. So much congestion causes traffic managers indigestion when they try to solve the problem. The desire of the passenger is to be picked up at the exact spot they exit the airport by their preferred mode of transportation. The goal of customer service minded rental car and hotel limo operations is to be at that specific area when the customer arrives or at least very close to the time the customer arrives at road side. There are complications for the rental car, taxi, and other limo services in that the airport authorities charge them for the number of times through the airport facilities and the dwell time. Dwell time means how long the vehicle is on the premises, and even includes various zones within the airport. The concept is that a vehicle is given a certain number of loops through the airport and can stay in any given zone for a certain amount of time before premium pricing takes effect.

The same space in front of the airport terminal is used for the taxis, shuttle busses, and limousines. Of course, there are all kinds of traffic flow schemes and facilities to handle these valuable traffic lanes. Arriving and departing passengers and the concept of public traffic lanes and public transit lanes are also part of the equation. Holding areas for taxis and limos are used to help control the number of vehicles in any one zone using the airport lanes.

'TAXI PLEASE'

As you arrive the at the terminal road side you see what looks like thousands of taxis parked in lots everywhere. Well at least there is the perception that this is the case. Actually, many airports do not give this perception any more. They use a holding lot, often off the airport premises to dispatch taxis. Some airports have multiple level parking garages for taxis. So even when you see a line of taxis there may be many others waiting a short trip away to fall in line when traveler demand warrants their use. The problem with some of the larger metropolitan airports is that the taxi pool is broken into territories with rules and regulations governing their operation. So certain taxi companies can operate in certain territories. Sometimes there is an overlap in the territories and for the most part customers don't care which company takes them to their destination.

TAXI CONTROL SYSTEMS

Taxi control systems have been around for a long time. They started with a line. Then tickets were introduced, and often, plain manual dispatch control was used. Some airports have taxi driver lounges and waiting areas so that the long wait can be in a better atmosphere. With long lines and even longer wait times there are problems like cutting into lines, and illegal passenger pick ups. If the radio frequency identification system is used for the driver and vehicle then applications like the loop and dwell time can be automated. Even before

FIGURE: AIRPORT TERMINAL APPLICATIONS

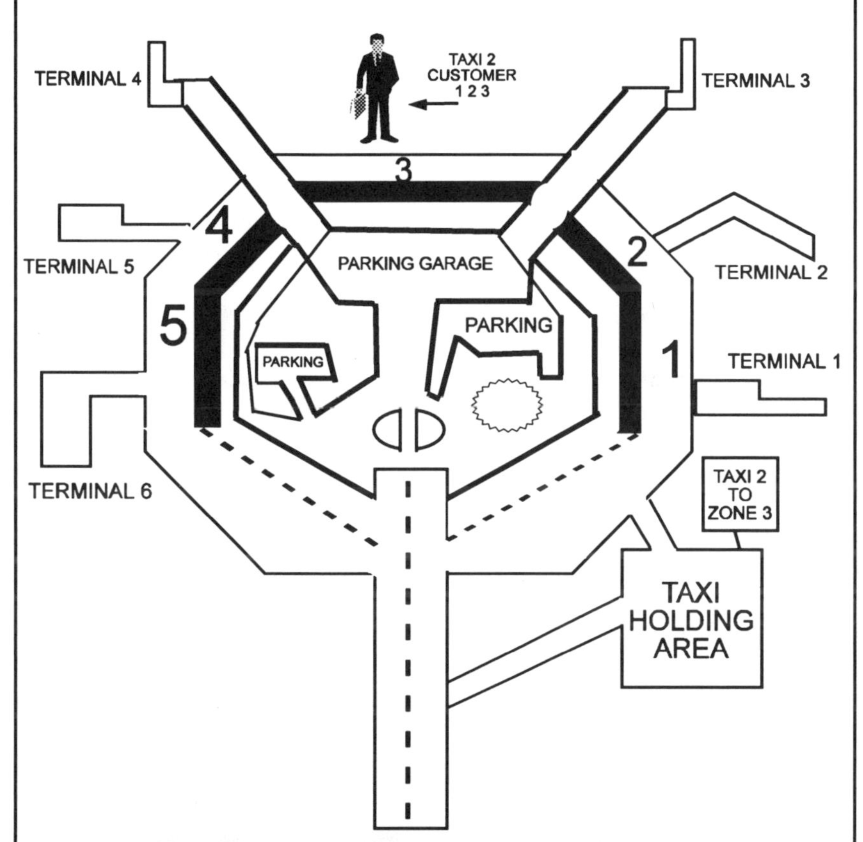

* MAXIMIZE TRAFFIC FLOW
* OPTIMIZE CURBSIDE UTILIZATION
* ENHANCE SAFETY & SECURITY
* INCREASE COST RECOVERY

the access to the passenger pick up roadway, there could be a priority scheduling dispatch system at the holding area parking lots. The drivers AVI tag could be identified, date and time information recorded, and the identification could then be associated with the proper queue. Readers placed strategically around the parking lots and access roads could help monitor the activity. Enforcement officers could randomly check vehicles using portable RF readers and communication links to the controlling computer. The data base could be checked automatically and the enforcement team could be notified of improper actions automatically. The passenger might go to a kiosk, and request a taxi for a given destination. If needed a ticket could be issued for both the passenger and the taxi. If the two tickets did not match, then the authorities could take the appropriate action.

HOTEL LIMO'S RENTAL CAR BUSSES

The traffic generated by hotel and rental car companies is also monitored. There may be usage fees based on loops and dwell as mentioned above. There is the aspect of customer service, so it behooves the company to have its shuttle where the customer is located. This service provides easy customer access, builds good will and in general helps the customer build a positive association with the company. On the other hand, to have a bus looping through the airport at too frequent an interval costs a lot of money. The concept of loop and dwell costs as well as driver time, fuel, and wear and tear is also significant.

Here another high tech concept can come into play. As the airport authority monitors the loop and dwell they could send a location message to the company. Instead of the company having two or three busses following each other without full loads, they could monitor the locations and dispatch additional busses based on more information. Inputs from kiosks could provide customer demand information in addition to the reservation systems.

AIR SIDE

The air side of the airport is generally thought to be secured. Security gates, guards, and metal detectors keep unauthorized people from entering this side of the airport. It is here that the airport authority and the airline companies must work together to provide on time arrivals and departures. The process can be very complex in the best of conditions. Once, while waiting for a flight I counted 47 different pieces of equipment that attended the plane on which I was to be a passenger. This included a fuel truck, a food truck, a cleaning crew vehicle, a crew bus, several carts full of luggage, tractors to pull the carts, vehicles for the maintenance personnel, unit load devices (containers for the aircraft), push back carts and on and on. I'm told that 47 is not the maximum. Questions such as, "What planes have been serviced?' and "What planes are ready for departure? are very important questions deserving immediate answers. Radio Frequency Identification could assist this process.

AIRCRAFT

The identification of the aircraft is designated on the tail with a unique code. While other even higher technologies might serve this function, the code could be part of an RF tag so that the aircraft could be identified at various locations in the airport. For example, a particular aircraft, at a particular gate, at a particular time, could be recorded automatically.

AIRCRAFT SERVICING

When a passenger or cargo plane is being prepared for a flight there could be a monitoring function performed that could provide significant benefits. As an example, all vehicles and service personnel could be identified by date and time in the proximity of the aircraft. The equipment and the employees and contract personnel could carry RF transponders. The Time Division Multiple Access, TDMA, protocol could be employed for surveillance.

SERVICE VEHICLES

If the various support vehicles were given radio frequency ID tags then significant operational data could be gathered and recorded. For example, the location of the particular push back carts, maintenance trucks, conveyors and the like could be maintained in a "where is" data base so that when exceptions need to be processed, locations for potential backup equipment could be identified quickly.

Food trucks, fuel trucks, and cleaning crew vehicles could also have RF tags for identification. Basic in and out-gate operations could be performed automatically. The locations of these vehicles could be monitored as they move throughout the airport.

UNIT LOAD DEVICES (ULD'S)

Unit load devices are the containers that fit into the aircraft. Particular aircraft can accommodate particular ULD's. When a particular type is required, it would be nice to know where the next available one is located.

Another concept that becomes important is accounting for the baggage and parcels that are loaded into the ULD. If, for example, a given ULD is assigned to the end of a particular baggage sort line, and the destination is known, then an audit can be made for the destinations of the baggage. When the ULD is loaded, an audit trail of identified baggage can follow it. The ULD can be checked when it is loaded onto an airplane to verify the destination, etc.

SECURITY

Security vehicles could be monitored as they travel throughout the airport. Particular routes and coverage by time could be analyzed to ensure that random checks are met and that patterns are not established without the knowledge of the security administrators. Gates could be controlled through the use of RF/ID. When the vehicle's, employee's or contractor's ID match the plan access would be permitted.

BUS/CREW/EMPLOYEE

Often employees and crew have off-site parking facilities, and busses are used to transport the employees to the secured side of the airport. This could be monitored if the vehicles had radio frequency tags. In fact radio frequency tags could be part of the employee identification. The identification could be made automatically without any disruption to movement. As the individual enters the bus a record could be made with the time stamp. If individuals without an ID entered, the system could alert the driver and security headquarters.

LUGGAGE CONTROL

Luggage control is a particular problem for the airport. There are many jokes about luggage arrival and baggage claim. My perception is that the airlines are doing a much better job than in the past. The use of bar code identification has improved the situation significantly. The concept of using a radio frequency tag for luggage has some appeal, because bar codes are optical. If the tag is not in view, there are missed reads and delays in handling the parcel. RF tags on the other hand can be applied and read without the optical line of site. These low frequency tags cost much more than the paper bar codes but there may be ways to reuse the tags. The tag would also help with the concept of positive boarding. According to this concept, if you do not fly on the plane neither will your luggage.

SUMMARY

The application of radio frequency identification systems to the airport and airline operations has significant potential. The increased productivity, utilization of resources, and traffic flow can certainly help. The frequencies of both high and low tags have a place in this environment, and the applications will continue to evolve.

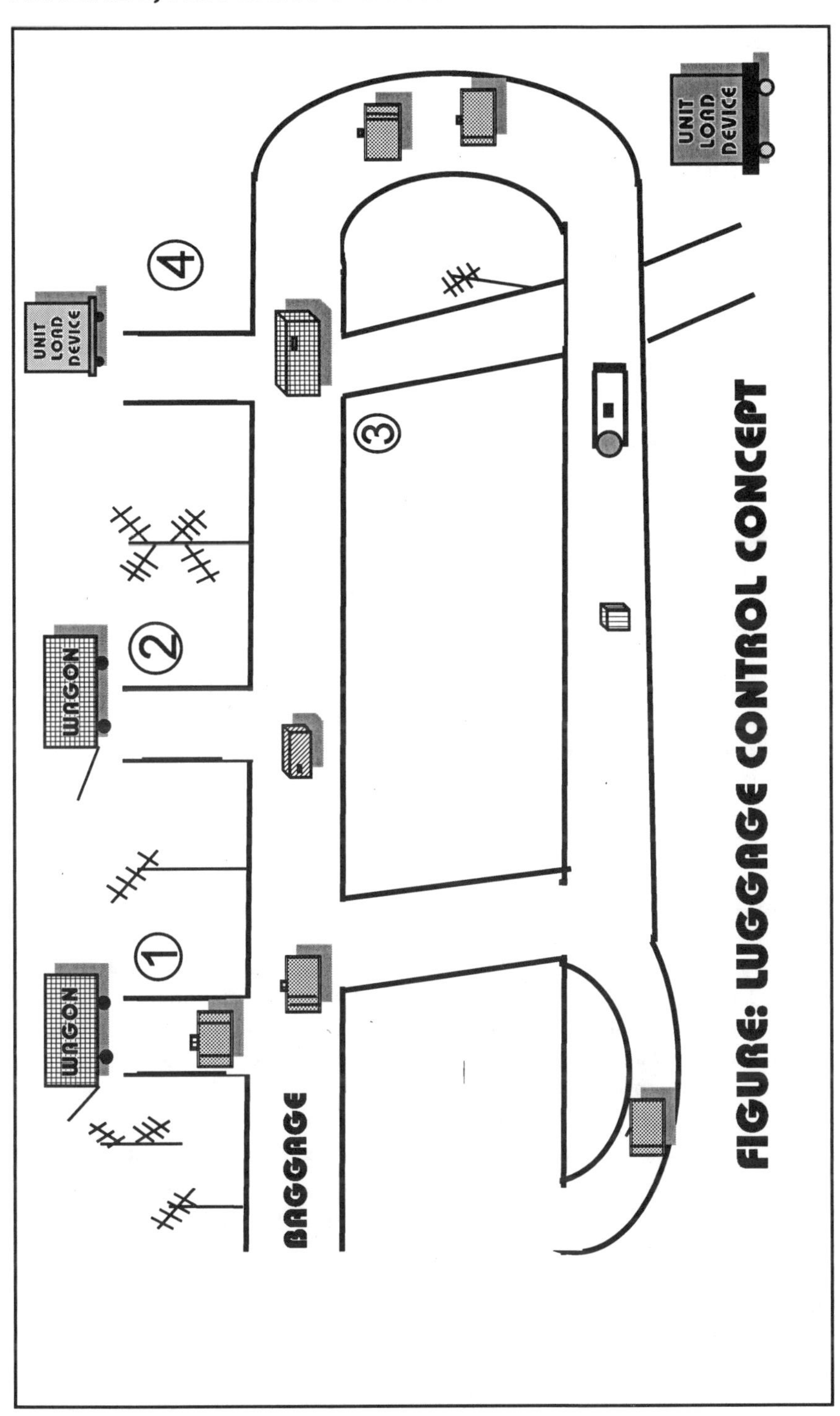

FIGURE: LUGGAGE CONTROL CONCEPT

CHAPTER XIV

RAILROADS

Railroads have big, bulky and expensive assets that are running around the various track systems of the world. Not only is RF/ID important for the identification of equipment but also from a control and maintenance view. AEI fits here.

In North America there are over 1.2 million rail cars and locomotives in service today. There is an equal amount in Europe and in Asia. These cars include box cars, refrigerated cars, hoppers, tank cars and others. The cars are owned by Class I railroads, other railroads, car companies and shippers. They travel across tracks owned by multiple railroads crossing state and often country boundaries. These assets cost thousands of dollars each. In fact, a locomotive might cost between $350,000 and $1.2 million. There are over 19 thousand locomotives in North America. Tracking these resources is critical to the success of the railroad operation. Information about the location of the asset and the contents of the asset is critical to the success of the railroad's customer.

A very strategic application identifies rail cars within a given railroad. The railroads have gone from a manual process of recording the car ID's to the use of bar codes and video systems. While two major attempts were made to use bar codes to identify the assets, environmental conditions prohibited timely and accurate identification. Dirt, ice, rain, snow fog all affected the readability of the optical bar codes. Special florescent stripes were used to make the bar code more legible to the scanners but the reliability was not high enough to produce the accuracy needed.

VIDEO SYSTEMS

Many railroads record the train on video as the train passes by a strategic point. There is often some degree of environmental conditioning required to produce reliable results. Lighted tunnels are built to insure that accurate images are captured. The process of handling the images can be very sophisticated. Consists lists are sent from the last point and a manual or visual comparison is made with the picture. Systems provide a best guess portion of the video to present a single picture for a data entry clerk to examine. When there is a discrepancy, the data entry clerk retrieves additional video pictures from the system to positively identify the rail car. The stenciling on each car is positioned by car type. Cars tend to be different lengths, and finding the stenciling is not a trivial problem. One of the key pieces of data is the car initial and number.

Key entry clerks view the video tapes, often with a computer assisted editor, and key enter the car initial and number. This is usually followed by a verification process. All of this care is taken so that accurate records can be maintained. While there are many applications that require identification, it is important to know that these applications feed systems like billing and settlement. In a sense the revenue health of the railroad lies in a balance of accuracy and timeliness of the identification systems.

It is not only the back office systems that matter in today's competitive world. Customers are requiring location information about their shipments. Just-in-time delivery is considered a must. The accuracy of the technologically assisted data entry systems is somewhere between 95 and 97 per cent. Fundamentally, the AAR's standard for AEI was born out of the desire to improve the accuracy to 99.9 percent.

Identification is not just important to North American. In Europe not only have the railroads piloted AEI applications for identification but they have set the pace for using transponders in the roadbed to assist in the control and usage of the tracks for high speed rail applications. Trains read and write to the transponders, so that subsequent trains will know time and date of the previous passage.

Providing high quality worldwide service to the railroad's (RR's) customers requires having access to the latest and most accurate information throughout the decision making process across the hierarchy of the railroad organization. Planning systems need the best available information. It is clear that RR's have chosen to use information processing technologies to achieve strategic business goals. The interdependence and integration of these advanced systems and technologies will help the RR's provide quality service to their customers.

TIMELY & ACCURATE DATA REQUIREMENTS

Important to the success of the RR's strategy is capturing data in a timely and accurate fashion. Data integrity is essential to making informed decisions. Automatic Equipment Identification plays an enormous role in this data capture process. And while radio frequency identification is at the heart of this data capture process there are other components of the system to be examined.

AEI SOLUTION

Fundamental to the Automatic Equipment Identification, (AEI), solution is the radio frequency system. Read-only and dynamic tags are available for the solution. Readers and antennas play an important role at strategically placed locations. With a network of AEI Systems spread along the track side, the systems and network management functions become fundamental to the operations on a daily basis. In addition, the distributed and host applications must be reviewed for the re-engineering of processes.

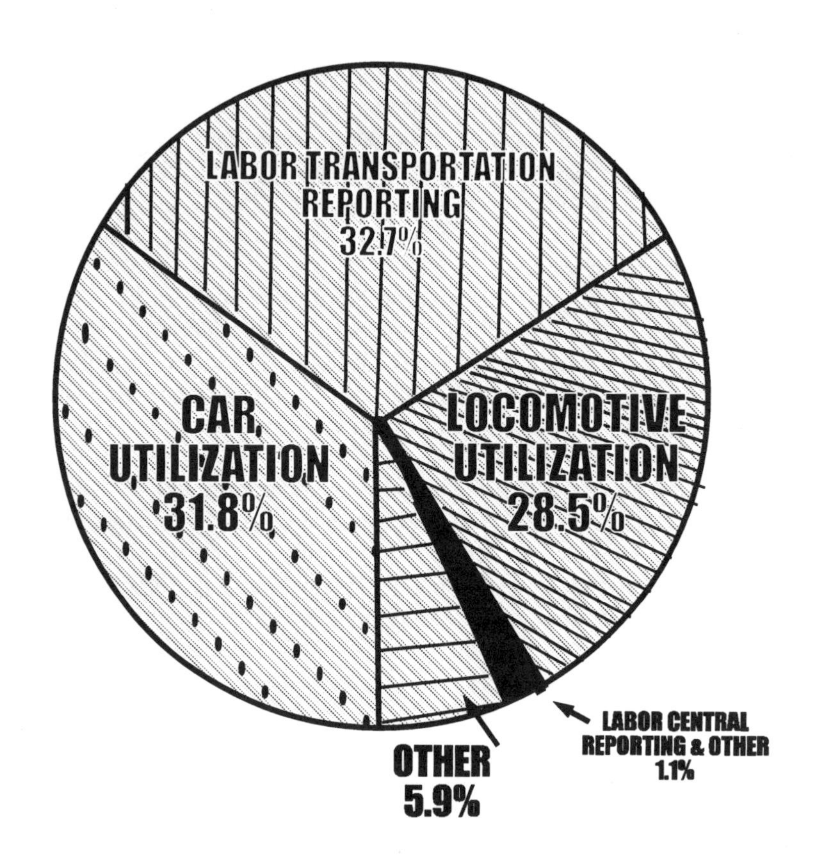

TOTAL ESTIMATED BENEFITS
$100.2 MILLION PER YEAR

SOURCE: O-T GENERAL COMMITTEE
AD HOC COMMITTEE ON AEI 5/91

CHART: AVERAGE ANNUAL BENEFITS - RAILROADS

READER EQUIPMENT

RAILROAD TRACKSIDE

PHOTOS BY BURNAM

OLD, NEW, REVOLUTIONARY SOLUTIONS

Once you consider the implementation of an AEI system, suddenly there is the ability to capture data at new positions within the system, and to capture other associated data. The cost factor for the new application is much less when the manual and often slow batch processing is out of the equation. Think of the possibilities of having a highly reliable identification of equipment and the ability to associate another related data point. Further, consider an automatic capture of the data and an immediate transfer of the data to the responsible management. The possibilities create a new demand for application systems and tend to place additional stress on the information processing departments.

While some of the applications will be discussed later in this chapter, it is important to provide an example or two here. Revenue systems need to know what cars pass by an interchange point. RF tags placed on the car could provide the tag ID, direction, and date and time stamp. Once the tag is applied, additional points of interest like hump yard and classification tracks might be of prime interest to yard management.

Further, marketing departments would like to know what cars are available for service on certain tracks and yards. If a fuel level indicator could be affixed to the dynamic tag, the ID and fuel level could be automatically sent to a power control system and a refueling decision could be made immediately. So even though the initial reason for AEI might have been car accounting, hump yards, marketing and power systems could use the new tool.

HYBRID RR SYSTEMS

The concept of hybrid AEI systems is important from a control point of view. If a reader could be set to read all of the tags that pass this point then there is a good chance that all of the tagged cars will be read. What about those with failed tags or those cars without tags? It might be well to use combinations of sensing devices in the data capture systems. For example some track side solutions use wheel detectors to count wheels and special routines to assist in the identification of the rail car. This hybrid identification approach increases the reliability of the data.

Systems may also require data inputs from acoustic wheel bearing detectors, and weigh scales. These systems when combined with the AEI system can be thought of as hybrid data collection systems.

WHAT DATA IS REQUIRED?

For any system to be effective it is very important to know the question being asked. An AEI system has the same concern.

1. What tags go by this point?

2. What rail cars go by this point?
3. How fast do the cars travel past this point?
4. How fast and in what direction do cars pass this point?
5. How fast, in what direction, and identify the leading end of the cars that travel pass this point?

The detail implied by each question significantly increases from question one through five. Care should be given to identify the proper question for the desired solution.

AEI RAIL APPLICATION INTRODUCTION

Before each solution in this chapter is described note that the emphasis here will be on track side data collection. Other AEI and AVI applications can apply to railroads that are not oriented to steel wheels on steel rails. These would be gate systems, motor freight, intermodal and drayage applications. These are covered in other parts of the book.

TRACKSIDE DATA COLLECTION SYSTEM CONCEPT

The Figure: Trackside Data Collection on page 157 shows conceptually the equipment that might be found in the Automatic Equipment Identification System. Antennas are placed on both sides of the track. The antennas are attached to reader systems and the necessary electronics inside the equipment enclosure. Lights and video equipment are found on some systems, depending on the questions to be answered. Wheel detector sets are a part of the configuration. Other collection systems such as hot box detectors and weigh in motion scales might also be found at track side. Included would be a communications link such as land lines or satellite communications. Of course, power and surge protection would be part of the system based on user requirements.

AEI APPLICATIONS

AEI Applications include those systems that need to be modified for AEI data collection. These systems are supported through manual data entry systems. There may be additional or new data collection systems and applications that will evolve over time. The following applications are provided as an introduction.

1. BILLING 2. CAR CONDITION 3. CONSIST 4. CONTENTS 5. FUEL STATUS
6. CONTAINER ON FLAT CAR 7. CUSTOMER NOTIFICATION 8. HAZARDOUS MATERIAL
9. CREW NOTIFICATION 10. HOT BOX DETECTION 11. KEY LOCATION NOTICE
12. LOCOMOTIVE TRACKING 13. TRACK AUTHORIZATION 14. YARD INVENTORY
15. TRAILER ON FLAT CAR 16. TRAIN BUILD PLAN 17. TRAIN SCHEDULING
18. WEIGHT & LOAD BALANCING 19. WEIGH SCALE SHIPMENT

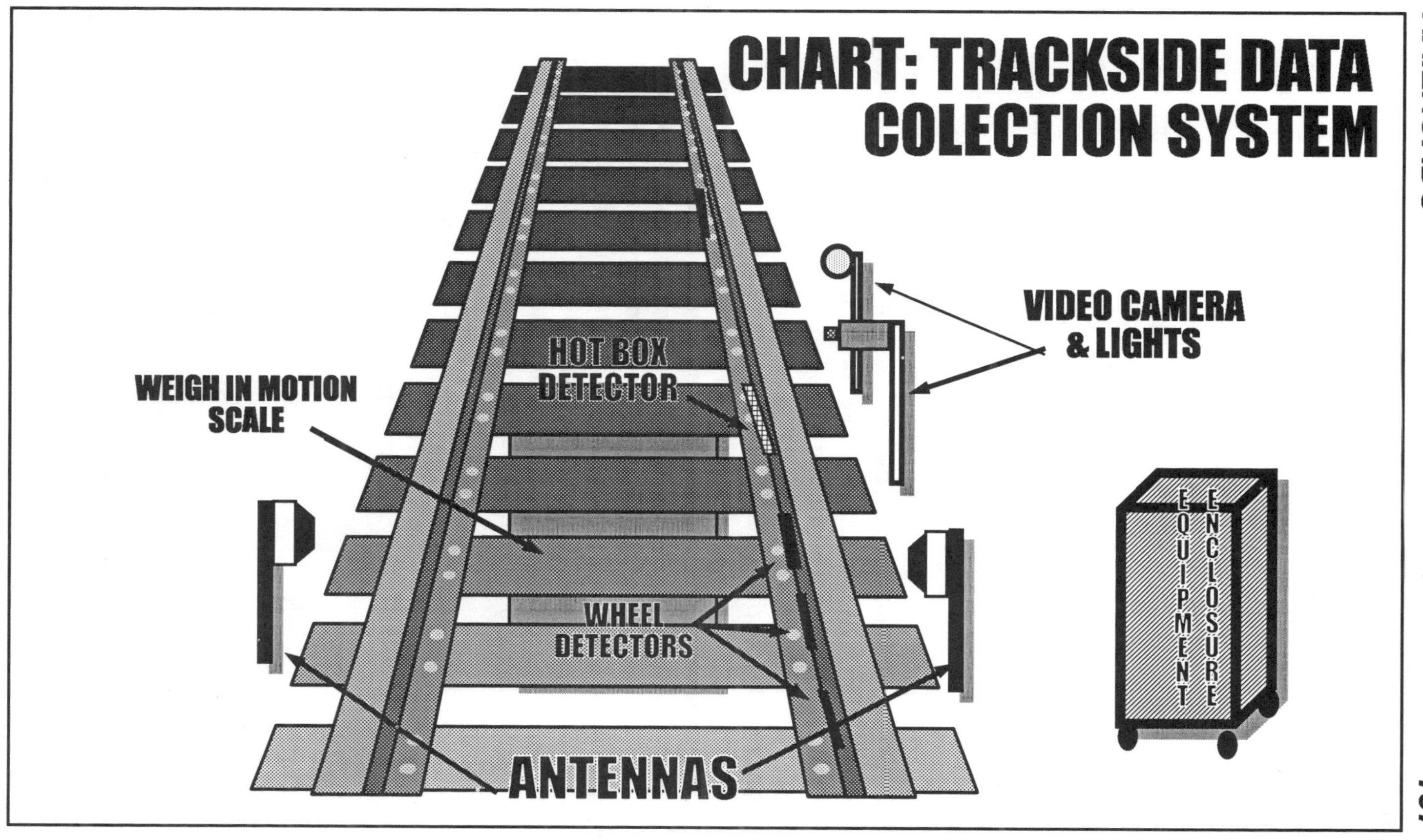
CHART: TRACKSIDE DATA
COLECTION SYSTEM
VIDEO CAMERA
& LIGHTS
HOT BOX
DETECTOR
WEIGH IN MOTION
SCALE
EQUIPMENT
ENCLOSURE
WHEEL
DETECTORS
ANTENNAS

BILLING / CAR ACCOUNTING

Shipments are contained in rolling stock (rail cars). Car movement records indicate the work that the railroad does on behave of the customer. Knowing when a piece of rolling stock is being used provides for the essential billing activity. With radio frequency technology, rolling stock usage history can be automatically collected and kept in history files that depict car movements. When a train departs or arrives at a yard, the movement to and from the mainline track can be recorded. Since the identification comes from a very reliable reader system and is automatically captured, the errors normally associated with manual entry would be eliminated.

With accurate and timely records the billing process should be more efficient. By integrating the RF applications, it is possible to inform the customer of the date, time, schedule, costs, anticipated delays or early arrival of their shipments. With this AEI system, it is possible to have billing that is closer to real-time, so that delays in receiving payment for services is reduced and the overall cash flow within the company is increased. Accurate records of car movements will help settle disputes and provide operation management with the identification of misdirected equipment.

CAR CONDITION SYSTEMS

In the operation of transporting the customer shipments from point A to point B, the condition of the car and the impact on the shipments are monitored. For those customer service options, special devices and data collectors travel with the car. Such sensors might be the impact and vibration detection equipment used to provide an indication of a quality ride. The use of a Dynamic tag could provide this collection. Reader sight could capture the necessary information to determine where on a trip the 'out-of-spec.' condition occurred.

Other sensors such as one that determines the height of the car and shipment could be provided. This is more likely to be a track side collector using AEI identification to positively ID a car that does not meet height restrictions.

CONSIST

The consist application generates a detailed record of the standing order of the rolling stock of a train. The consist or "clean list" is the heart of the train equipment movement. The consist of the train changes at the yard and at rail sidings when a car or block of cars are cut from or added to the train. The consist is required to make up a train; to verify that the train was made up correctly after leaving a yard; to identify any changes in a train before entering a yard; and as input for car classification when entering a yard.

The consist records may include the following information:

* Locomotive and railcar type
* Car initial and number of the rolling stock
* Fill records for unidentified equipment
* Date and time of train passage
* Speed and direction of the train
* Count of locomotives and railcars

The method of creating the consist without RF/ID is a manual process. While system generated lists create the initial consist reports, manual intervention is utilized to perform all other aspects of the consists. The standing order of the cars in the train is important to operational planning systems, to optimize movements and control. The objective is for the system to have a record of what is, rather than forcing reality to match what the system had planned. For example, if the system thinks that car A is in a block of cars 123, but block 123 does not contain car A, then operations should know this. They should be able to check whether car A traveled with block 123 to another destination when in fact car A may have remained in the yard. So the objective is to know what is real.

Radio frequency tags placed on two sides of each car help control the movement of the car according to plan, and automatically record any exceptions to the plan. RF readers are placed at strategic locations within the rail system. With AEI readers it is possible to automatically identify the cars located on classification tracks and the arrival and departure tracks and the main line tracks. The clean list can be built from this information.

CLEAN LIST

To show how the clean list application can improve day to day operations consider the following chart showing how the previous methods compare to the process with RF/ID tags. The information can be made available within seconds using automatic equipment identification, AEI, as compared to minutes, hours or even days in the old methods.

CHART: CONSIST/CLEAN LIST

TRADITIONAL	VIDEO TAPE	AEI
1. WALK OR DRIVE	1. VIDEO TAPE	1. READ RF ID'S
2. WRITE ON PAPER	2. TRANSFER OR TRANSMIT TO CENTRAL	2. TRANSMIT TO SYSTEM
3. TAKE LIST TO OFFICE		
4. DATE & TIMESTAMP	3. SCHEDULE BATCH	
5. SCHEDULE BATCH ENTRY	4. DATE & TIMESTAMP	
6. KEY ENTER ID	5. KEY ENTER ID	
7. VERIFY	6. VERIFY	
8. TO SYSTEM	7. TO SYSTEM	
MANUAL	**MANUAL**	**AUTOMATIC**
MINUTES - HOURS	**MINUTES - HOURS**	**SECONDS - MINUTES**

The AEI systems automatically identify the passage and direction of a car that passes a reader position. In the process of building a block of cars it is often required that the cars will move back and forth along the track. The automatic system will record each movement. The final configuration is the one we want to process in the computer system. The data collection system used at trackside must coordinate with the rest of the system to account for multiple interim movements.

The Table: Automatic Car Movement is an example showing the concept of this movement.

A	B	C	D
T 1 WEST	T 1 WEST	T 1 WEST	T 1 EAST
1:06 PM	1:14 PM	1:28 PM	1:45 PM
LOC A	LOC A	LOC A	LOC A
CAR 1	CAR 1	CAR 1	CAR 1
CAR 2	CAR 2	CAR 2	CAR 2
CAR 3		CAR A	CAR A
CAR 4			CAR 3
			CAR 4

TABLE: AUTOMATIC CAR MOVEMENT

Column D is the final configuration in the above example but column A, B, and C would also be generated.

In addition, it is not uncommon to cut a car from a rail siding into a block of cars. Such movements can also be recorded with strategically placed readers. When the activity of this type is high, a reader or two may be justified. However, when the activity does not justify a reader, other mainline readers would pick up the inserted car in the consist report. Routines to determine split blocks and cars that are remaining in a string or a standing order would assist in the understanding of what is new to the train. Keep in mind that the action might be to cut a car from the train or to add on to it.

The consist application can automatically record all car movements in a timely and accurate fashion. This application is really the bread and butter application that required AEI in the first place.

CONTENTS

The contents application area is not yet fully explored by the standards groups but the concept is one that is fundamentally sound customer service. The initial mandate for car tagging is for the read-only tag to be placed on the cars. Read-only tags are not intended for contents applications. The dynamic tag, and even more likely, the read-write tag technology is. Industrial users could write the bill of lading or an electronic way bill onto a read-write transponder affixed to the side of the car. This can save millions of dollars for the industry and provide valuable time saving access for government authorities.

CONTAINER ON FLAT CAR

The use of AEI equipment for this application has significant benefits for the customer, operational personnel and train management. The benefits manifest themselves in many ways because of the requirement to know and understand the identification of so many pieces of equipment in the intermodal process. These include, but are not limited to, the rail cars, containers, refrigeration units, and lift facilities. The planning function is very complex. Container types are different sizes. Stacking options vary depending on weight, content and size and there are differences in the types of rail cars that are used and the facility at the destination that need to be matched. With all of this complexity caused by seemingly limitless planning combinations and restrictions there has never been a more obvious need for a toll like AEI. Here are some of the uses of AEI in the COFC environment.

1. Identify the rail cars, the standing order in the block.
2. Identification of container and container type.
3. Chassis identification.
4. Other equipment ID.

These inputs can provide planning systems the information they need to optimize operations and meet customer demands.

CREW SCHEDULING

Labor is an expensive part of doing business. The government has requirements concerning the continuous hours worked by train crews. Insurance coverage, union rules and restrictions, make it important to utilize employees in a safe and efficient way. Using RF technology to determine the departure time of a train and the expected arrival times of trains, train crews can be scheduled on a timely basis. This information tends to help optimize the utilization. From time to time there are unforeseen delays that cause schedule changes. The AEI system can provide input of the status of the train as it moves through the system. Crew staffing can be optimized for the train, reducing problems with over and under staffing. It is also possible to schedule crews based upon layovers, days on the road, and destination.

CUSTOMER NOTIFICATION

Time and time again, customer satisfaction surveys maintain that they need information about their shipments. Many believe that the information is as valuable as the shipment itself. The AEI system can assist in providing the information in real time.

If readers are placed at predetermined customer strategic locations, the information about a passing car could be recorded. This record could be sent

ACOUSTIC WHEEL BEARING DETECTION

immediately to the customer. A simple table scheme could insure that the right car information is being sent. In this way status information could reach the customer and corrective actions could be taken in time. This could apply to just-in-time applications and simple alert applications.

As an example, the receiving customer yard requires a notice to their crew one hour in advance but because of scheduling variables, the arrival time of the train was not consistent. If a reader was placed away from the receiving yard, say one hour and fifteen minutes away, then the management would have a fifteen minute window to call the crew.

HAZARDOUS MATERIAL

There are so many applications and procedures for hazardous material. The processes for the equipment like cleaning actions and interactions with various officials, routing decisions, movement plans need to be recorded. If the AEI system used the dynamic tag or the read-write tag, then there are ways to control the hazmat information flow. Again, information can be used for planning systems, operational management in the real time, actions taken by safety crews at inspection points and the dreaded scenes of accidents.

HOT BOX DETECTION

One of the hybrid systems that can provide a great benefit to the railroads is the hot box detector system and AEI. The acoustic wheel bearing detection systems and heat sensing systems have been applied in the field for a long time prior to AEI. The application was a wheel set count along with an out of specification indication from the detector would indicate a problem with a train. If the problem is severe enough, the train would be stopped and a crew would count axles until the defective wheel set was identified. Counting wheel sets from the beginning of a train presents many problems. Often the wheel set would not show the problem in a physical way. If the crew was off by one or two, the real problem might not be fixed. As well intended as the crew might have been, the environmental conditions and human error would influence the corrective action. AEI applied to and integrated with the hot box detector could specifically identify the car, and with the proper system design, could identify the end of the car with the problem. This could provide a significant advantage and a safety improvement.

FUEL STATUS

This application involves the use of a dynamic tag and a connection to the fuel gages of the locomotive. The idea is a simple one. When a locomotive arrives at a yard, a reader, strategically placed, reads the locomotive ID and the fuel gage. If there is enough fuel, and other conditions are met, the locomotive can be available for service immediately. Today the locomotive pulls the train into the arrival track, and then goes to a refueling station, sometimes to be topped off. Power decisions can be made by knowing the locomotive is immediately placed in the ready queue.

There are other applications that this fuel level detection system can support. Even though the practice of sharing locomotives is not as common, the ability to have better control of fuel costs, and shared usage times, may affect some of the thinking in this area. Fuel costs are a significant part of the cost equation so anything that might help the understanding of how to share the resource could provide customer service benefits and provide some opportunity for a competitive edge.

The other benefit from fuel level indications is the automatic collection of fuel usage. If the locomotives in a train are not consuming the fuel in a relatively equal basis then there is cause to examine the details of the pull. This may be a flag to maintenance that something is wrong. More timely and accurate data makes for more informed decisions.

KEY LOCATION NOTICE

As AEI reader systems are positioned throughout the track network, the concept of using the reader for more than mere reporting is possible. If strategically placed readers are located at critical points, then operation management and personnel can be automatically notified when a train passes the point. This concept might work for special requests like, "If this train reaches this point by 3:48 PM then notify the yard master."

LOCOMOTIVE TRACKING

Locomotive tracking or power distribution applications need to have accurate data in order to be effective. When managing a fleet of locomotives, there is a tendency to have these power units migrate to the same locations within the system. It is really more complicated than this but, as an example, it takes more locomotive power to pull a fully loaded train of coal cars, than it does to pull empty coal cars. So power distribution systems have been looked at as helping to solve the problem of having the power units migrate to one end of the system. AEI systems could provide accurate and timely input to these systems.

TRACK AUTHORIZATION

Trains travel across segments of track that are monitored by control centers. Strategic placement of RF readers could provide data relative to the date, time, speed and direction of the locomotives, rail cars and end-of-train devices. This information could be sent to the control center for a graphic display of all the trains in an area. This data would supply the control center with additional information on the status of the train. In addition, other equipment like service vehicles could be identified. This would help with collision avoidance. Using the AEI systems in this way could provide a secondary source or redundant information for track authorization. This kind of operational system provides a way to increase utilization and maintain safety.

TRAILER ON FLAT CAR

This application is similar to that of the COFC application mentioned earlier. The Trailer on Flat Car, or TOFC, application is a wheeled operation that involves a trailer or a chassis container combination sitting on a special rail car. There are special destination facilities concerned here. Certain rail car types might not fit the physical facilities with given lift equipment and so forth. AEI will help provide accurate identification of the equipment being used, so that planning systems can assist in the decision support area.

TRAIN BUILD PLAN

Building a train is a very complex task. There are many decisions that need to be made. Typically, there are hundreds of rail cars, each with a specific destination. The concept is to sort the cars into blocks of cars with the same relative destination, but there are factors and characteristics like weight, height, and width of each car that can certainly affect the routing of the train and the number of power units required to pull it. The information needs are significant to say the least.

Classification tracks are used to assist the sorting process. As cars are identified for a particular block of cars a designated classification track is used to store the car prior to the final train building process. Based on available space, the track classification designation may change from one to another. For simplicity sake, say a given track has Chicago as a designation in the morning, and Columbus at night. The trick is to empty the track for the use of the next shift or for the next train.

The information needed in this operation comes from different sources. Keep in mind that the characteristics of the car is important. Certainly a railcar length restriction would prevent the car from traveling to a destination. For now consider the following sources of cars.

A. Yard Inventory
B. Incoming consists
C. Local pickups
B. New Bookings

Consider that the schedules and actual arrivals and departures tend to change or be different based on daily operation, the problem is mind boggling. What about meeting customer service agreements?

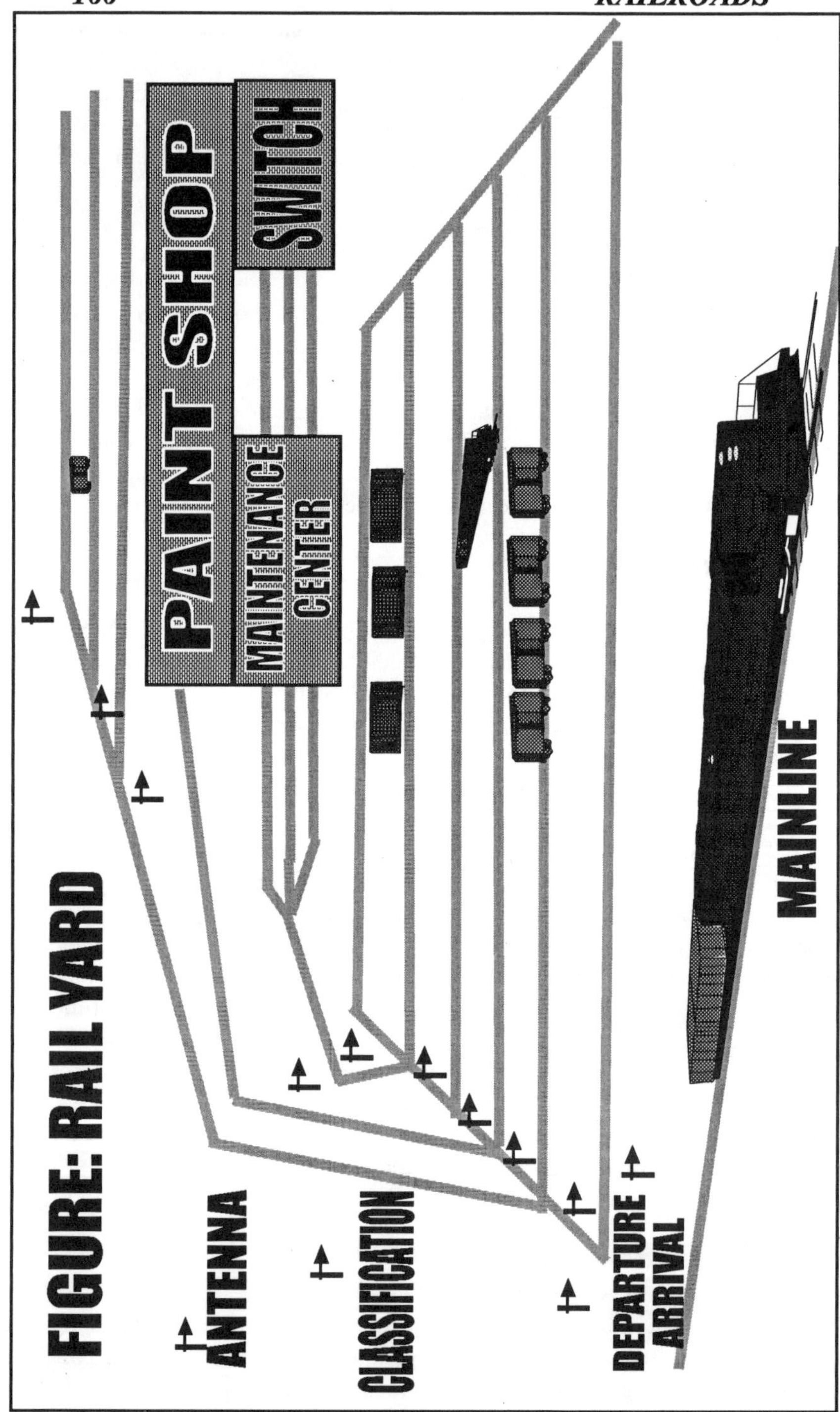
FIGURE: RAIL YARD
ANTENNA
CLASSIFICATION
DEPARTURE
ARRIVAL
MAINLINE
PAINT SHOP
SWITCH
MAINTENANCE CENTER

If antennas and readers supply automatic identification of the RF transponder, then an accurate data base can provide planning systems and operational control systems and the necessary real time data to make adjustments. For example, in a hump yard, cars are pushed over the hump and allowed to roll down hill to be switched onto a classification track. The AEI system could provide arrival and departure information for immediate decisions. Errors can be detected and potentially corrected before the train departs.

TRAIN SCHEDULING

The railroads make money moving commodities and shipments from here to there. Customers need to know that their products can be moved from one location to another on a reasonable time schedule. Customers have often listed consistency of service as very important. The 'no surprises' philosophy applies here. To meet customer needs, railroads have time schedules that must be maintained for customer satisfaction.

In order to be able to transport customer goods, rolling stock must be available at a time and location that can satisfy both the customer and the railroad. Train scheduling applications are constructed to enable both the railroad and the customer to meet their goals. The application receives its input from the marketing application in order to know what equipment and commodities need to be transported. Yard applications and others are queried for existing equipment and the consist application to know what equipment is inbound. These systems can be used to assist the building of the train and the schedule required.

The AEI systems feed the data to the applications, and when a particular car is behind schedule, the reader systems are instructed to search for particular cars and notify operations when they are found. In this way an immediate reaction to the problem helps establish an on- time delivery.

WEIGHT AND LOAD BALANCING

Trains are dynamic in that they change configurations, start and stop, negotiate curves and inclines, and traverse many different types of track. This requires that a train be properly balanced and not overloaded. Combined with static and dynamic car weighing systems, RF/ID technology is used to identify, specifically, the rolling stock that is being weighed. The tag information contains tare weight, car type and the like. Combining the identification with the actual weight of the car in its current configuration will provide needed information to the system, such as work orders. Speeds and breaking calculations can be made.

WEIGH SCALE SHIPMENT

In the bulk applications it is the weight of the shipment that matters most. It is not just car loads but how much product is being delivered or hauled. Cars can be weighed to determine the before and after differentials. The tare weight, empty weight, loaded car weight and the new empty weight can now be made available automatically. This offers particular application opportunities to insure customer service and to manage customer contracts. This application can be implemented with read-only technology, but has even more possibilities with a read - write transponder. For example, memory could maintain several sets of before and after weights with time stamp information, scale used, etc.

YARD INVENTORY

The yard inventory application contains detailed information pertaining to the type, identification and location of the rolling stock within a yard. Utilizing this information enables yard managers to place equipment for speedier train make-up, to group equipment and form blocks of cars, and to schedule maintenance.

If you view the diagram on page 166, showing a rail yard, consider points for placing RF readers and their antennas. The entry and exit points are good examples of this. The placement of readers would provide data to a system showing what cars are on the track. Even servicing station and maintenance bays could have readers that show which particular car or locomotive occupies a maintenance position.

SUMMARY

These applications at track side provide significant opportunities to improve operations and save time and money. Automatic Equipment Identification, AEI, sets the stage for significant benefits for the railroad and their customers.

CHAPTER XV

INTERMODAL

Intermodal applications of RF/ID center around the identification of the container. Containers are sent all over the world. The identification of the intermodal equipment offers significant opportunities. The expanding capabilities of RF/ID offer significant potential improvements in tracking and handling.

The industry has seen an increase in intermodal shipping activity. The term container means more today than ever before. A few years ago it would have been taboo to mention truck and train in the same conversation, but now we drink coffee and eat dinner at the same conferences. Ocean shipping company representatives share the podiums with railroad and trucking companies. They all invite drayage representatives to the table and common problems are discussed. Of course the shipping community is looking at a world wide picture with new trade agreements being signed and strong competition. New concepts are being tried to increase productivity and lower operational costs. Railroads claim an accuracy of 95% for on-time delivery. Truckers use rail for long hauls. Ocean carriers set agreements with both. The industry holds seminars assigns committees for the advancement of new technologies. There is significant opportunity for everyone in this marketplace.

Intermodal Marketing Companies participate as well. Some are advocating a name change to Logistics Management Companies. Why not? There are so many players and so many modes. Here are a few:

1. Container-On-Flat-Car
2. Trailer-On-Flat-Car
3. Container-On-Barge
4. Trailer-On-Barge
5. Cargo By Air
6. Ocean Shipping
7. Less than Truckload
8. Truckload

There are so many options for shippers and IMO's, it is no wonder that the industry is calling for more efficient ways to operate within this logistics chain. Shared resources make it a requirement that better tracking and controls be placed on the equipment. There is so much equipment.

Containers, chassis, gen sets, reefers, and tractors, not to mention lift trucks, stackers, cranes, spreaders, and straddle carriers, etc., have been estimated to be in excess of 6 million units. Many think that estimate is low. This equipment is expensive and much of it is shared in chassis and container pools. Much effort is used to capture the ID of the containers in the process. Those in the ocean shipping business insist that 100,000 times a day a long shoreman writes the ID on a soggy piece of paper to be key entered into the system and the process is 85% accurate. It is no wonder that automatic equipment

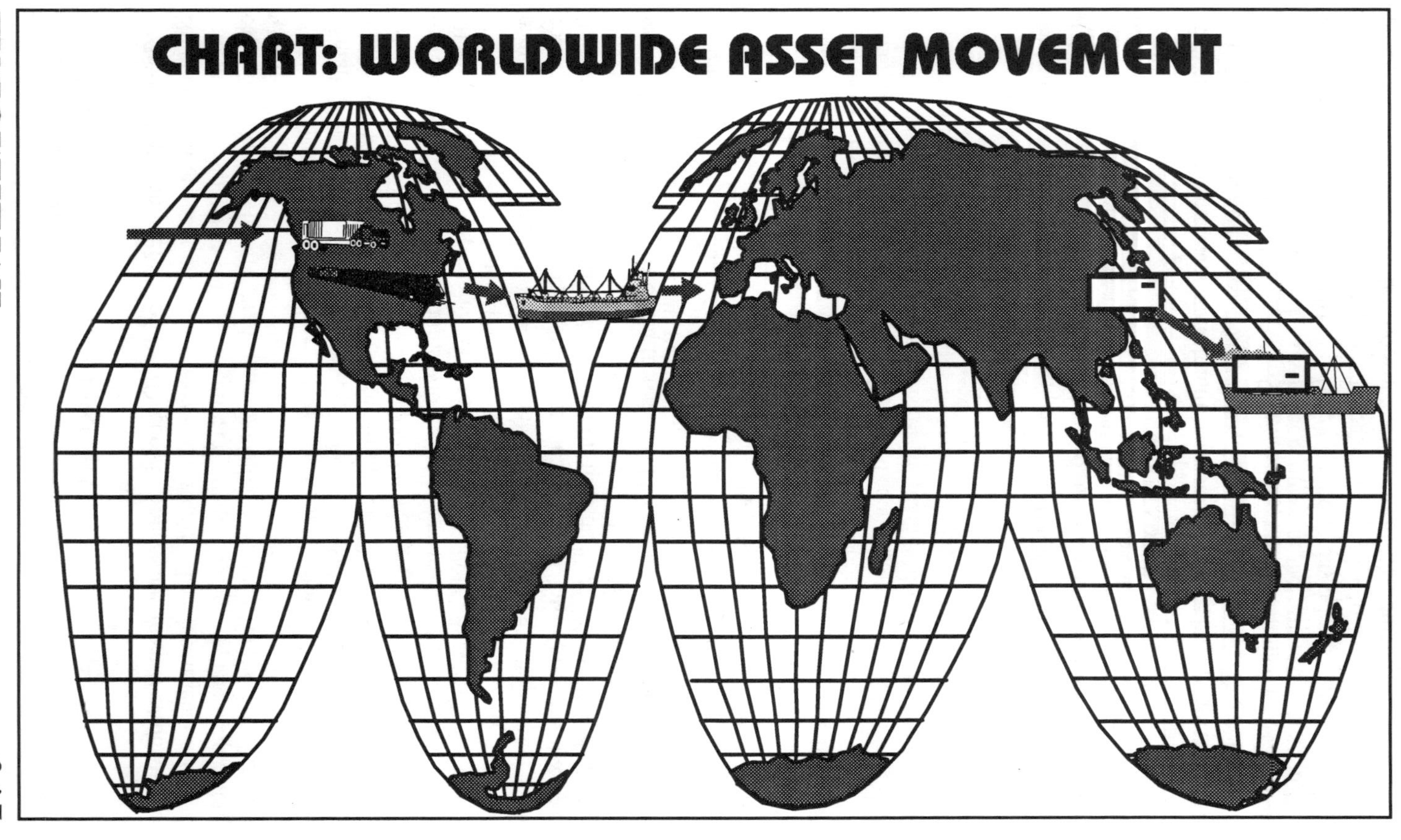
CHART: WORLDWIDE ASSET MOVEMENT

identification is one of the hottest topics. In fact, one ocean carrier committed $15 million in 1993, to tag their container fleet.

WORLDWIDE TRANSPORT

Multiple modes of transportation are used to ship goods from one part of the world to another. A container at a shipper's dock is filled with goods. A drayman picks up the container and takes it to an ocean port. Inspections and customs processing takes place. Stevedores transport the container to the crane for stowage on a ship. Planning systems insure that the ship is loaded in an optimum way. Load and weight balance are taken into account. The ship arrives in port and is discharged. A transtainer stores the container in the yard. It is placed based on an intermodal rail move that is scheduled for two days from now. Move paper is processed. Along the route a system reads the container ID and reports the progress to the destination.

Even with the short description given above for container movement, it is easy to see that there are many different organizations, work groups and authorities who are part of the process. There are leasing companies for containers and chassis and even special equipment leasing. There are separate repair companies for the containers. With all of this there is a significant need for information throughout this process.

INTERMODAL EQUIPMENT

There are many players in the system to transport a container and the relationship and information flow needs a strong system support. The types of equipment used in this process must be optimized for efficiency. Understanding where equipment is located becomes important to the process. Some of the equipment can assist us in the data collection and tracking of container movements. The equipment used includes the containers and other more sophisticated tools. Here is a list of some of them.

A. CONTAINER	I. GANTRIES
B. TRAILER	J. YARD CRANES
C. RAIL CAR	K. SPREADERS
D. TANK CONTAINER	L. REACH STACKERS
E. REEFER	M. STRADDLE CARRIERS
F. SEALS	N. FLT
G. CHASSIS	O. HOSTLER/TERMINAL TRACTORS
H. CHASSIS STACKING SYSTEM	P. TRACTORS

There are containers to carry the shipments, other things to carry the containers for long hauls, and lift equipment to move the container from mode to mode. The pieces of equipment that have motors and power are becoming more sophisticated. They have electronics that not only tell the health of the

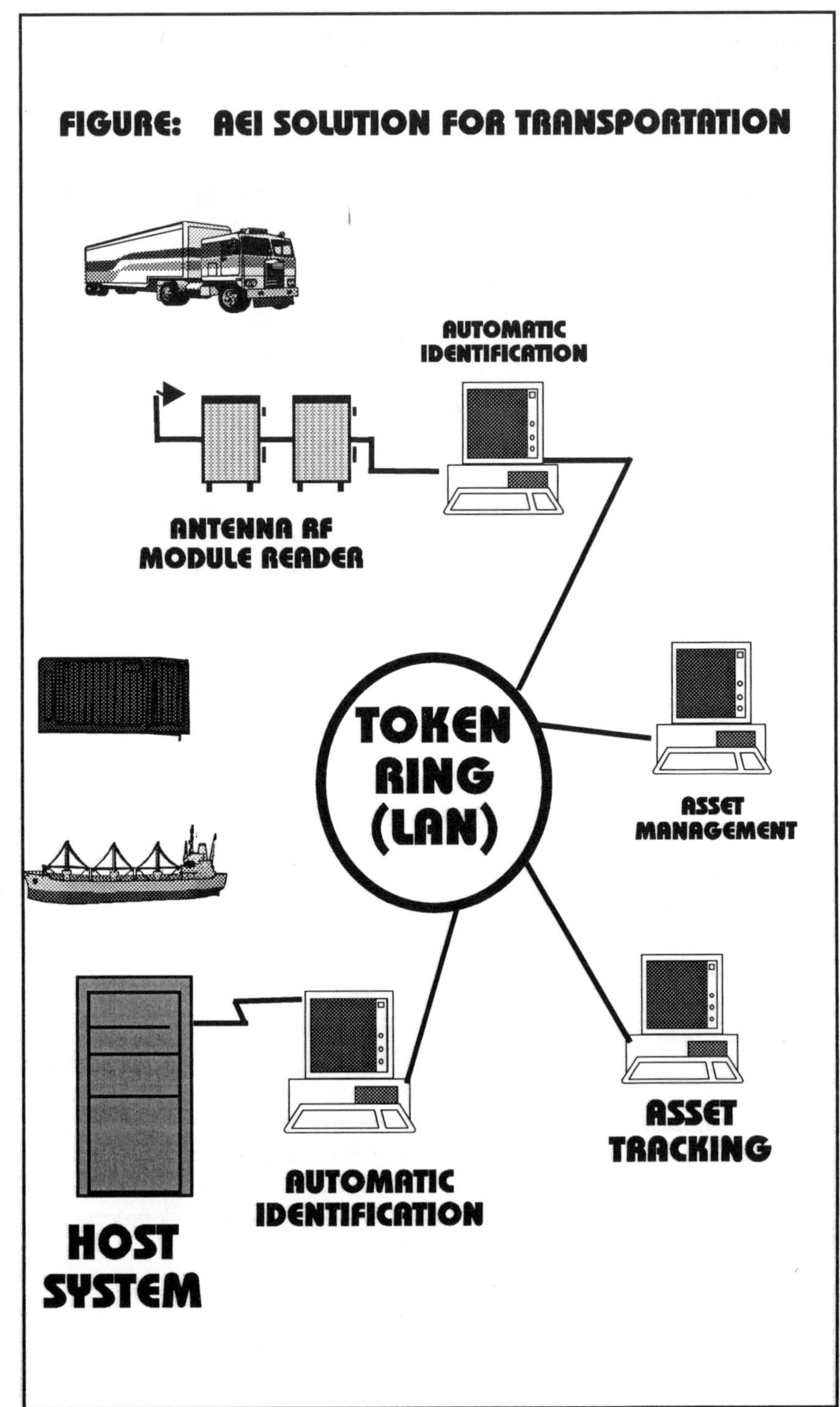
FIGURE: AEI SOLUTION FOR TRANSPORTATION
AUTOMATIC IDENTIFICATION
ANTENNA RF MODULE READER
TOKEN RING (LAN)
ASSET MANAGEMENT
ASSET TRACKING
AUTOMATIC IDENTIFICATION
HOST SYSTEM

unit itself, but also can provide certain operational data. As an example, whether a spreader bar is open or closed, and the height of the lift can be captured. Key entry and even voice entry systems have been tried to capture the identification codes of the container.

TECHNOLOGY IMPLEMENTATION

There has been significant research and development (R&D) money spent to research the use of radio frequency identification systems. Since containers move around the world and the identification of the container is the issue, it is important to have a technology that can be used around the world. The standards groups continue to work on these issues, but for now, let us examine the use of RF/ID.

Technology tends to be employed first within the framework of current applications and then expands into new uses. For example, if the gate process includes the identification of a container, then we might employ an RF/ID system to provide an accurate and timely identification of the container. Once the RF transponder is applied to the container, and the system is installed to read the ID at the gate, we might enhance the gate processor system by asking it to notify the maintenance department when container ID 123000 arrives at the gate. So an additional function is added to the initial system.

RE-ENGINEERING

While incremental improvements can be made by the modification and integration of new technologies into existing systems, there are very significant opportunities if a re-engineering of processes can be implemented. When the entire organization assesses its ability to capture data and make it available, and then sets a process in place, radical changes can be made in a reasonable order. This is risky but the rewards are great. The risk can be contained with planning and organizational influence.

RF/ID APPLICATION INTRO

There are many situations in the process of handling the container where RF/ID might be of significant assistance. Some might be as simple as a toll road application or an IVHS system. To provide a basis for discussion, there are five areas to be addressed: Gate - Yard - Lift - Rail - Management.

The idea is to address what might be identified and how the process might work. It is a way to indicate where the ID is important.

GATE

The gate has been a topic for improvement for a long time. Particular stress is

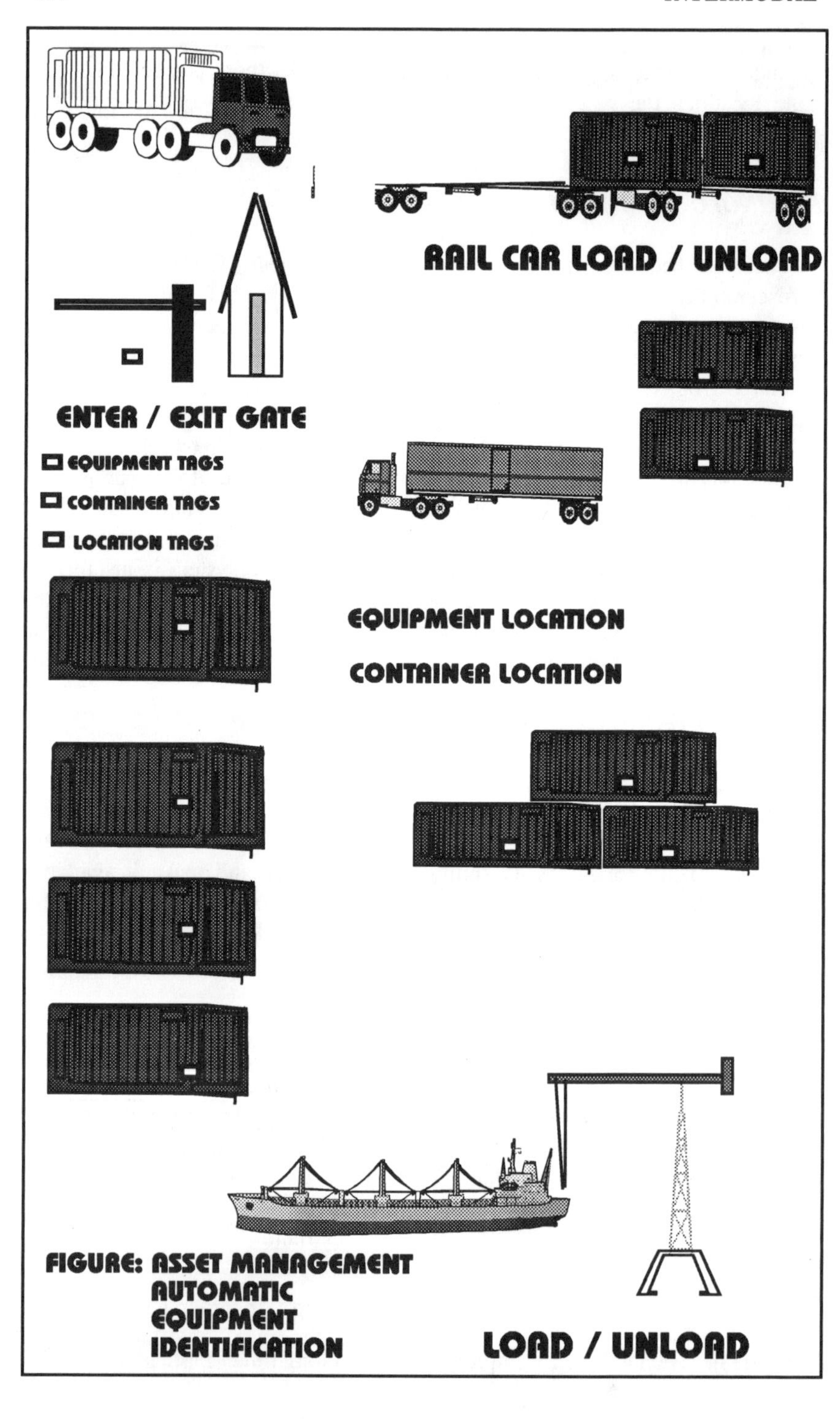

FIGURE: ASSET MANAGEMENT AUTOMATIC EQUIPMENT IDENTIFICATION

placed on the operations at ocean terminals. Land is not always available to expand the number of gates, and, when it is available, there is a cost factor. In some locations the gate processing is so slow in relation to the amount of traffic, that there are tractor trailers waiting in lines long enough to block freeways. So, gate systems that decrease service time through the gate becomes a paramount concern.

If we consider the gate from an equipment point of view there are a variety of configurations.

EQUIPMENT	BOBTAIL	SINGLE	DOUBLE	TRIPLE
TRACTOR	X X	X	X	X
CONTAINER		X	X	X
CHASSIS	X	X	X	X
CONTAINER			X	X
CHASSIS			X	X
CONTAINER				X
CHASSIS				X

CHART: EQUIPMENT AT A GATE

As you can tell from the chart above, there are a number of combinations for identifying the equipment through the gate, and there may be more. If gen sets are used, they also could be identified. Other processing options and technologies are available. Let's examine some of the scenarios that might be possible with RF/ID.

BOBTAIL GATE

The gate setup for an intermodal or container yard might be dedicated to handle a tractor without a chassis and container or even a tractor with a only a chassis. There are no container movement through this gate. RF/ID tags could be placed on the equipment or even given to the driver when picking up a container. The identification can be made and the system could produce the paperwork while the entry is made.

If there is an Electronic Data Interchange, (EDI), system to forewarn the gate system of the pending arrival, paperwork could be ready. A simple location ticket could be printed giving a driver the pickup location, or if traffic volumes justify, an area could be set aside for such pickups.

OTHER RF/ID GATES

In the Bobtail gate example above there was very little interaction with the driver required until the point of physical pickup. In other gate situations, there is a lot of interaction required. After visiting many gates at both ocean ports and rail intermodal sites inland, a better appreciation of the problem

FIGURE: IN / OUT GATE

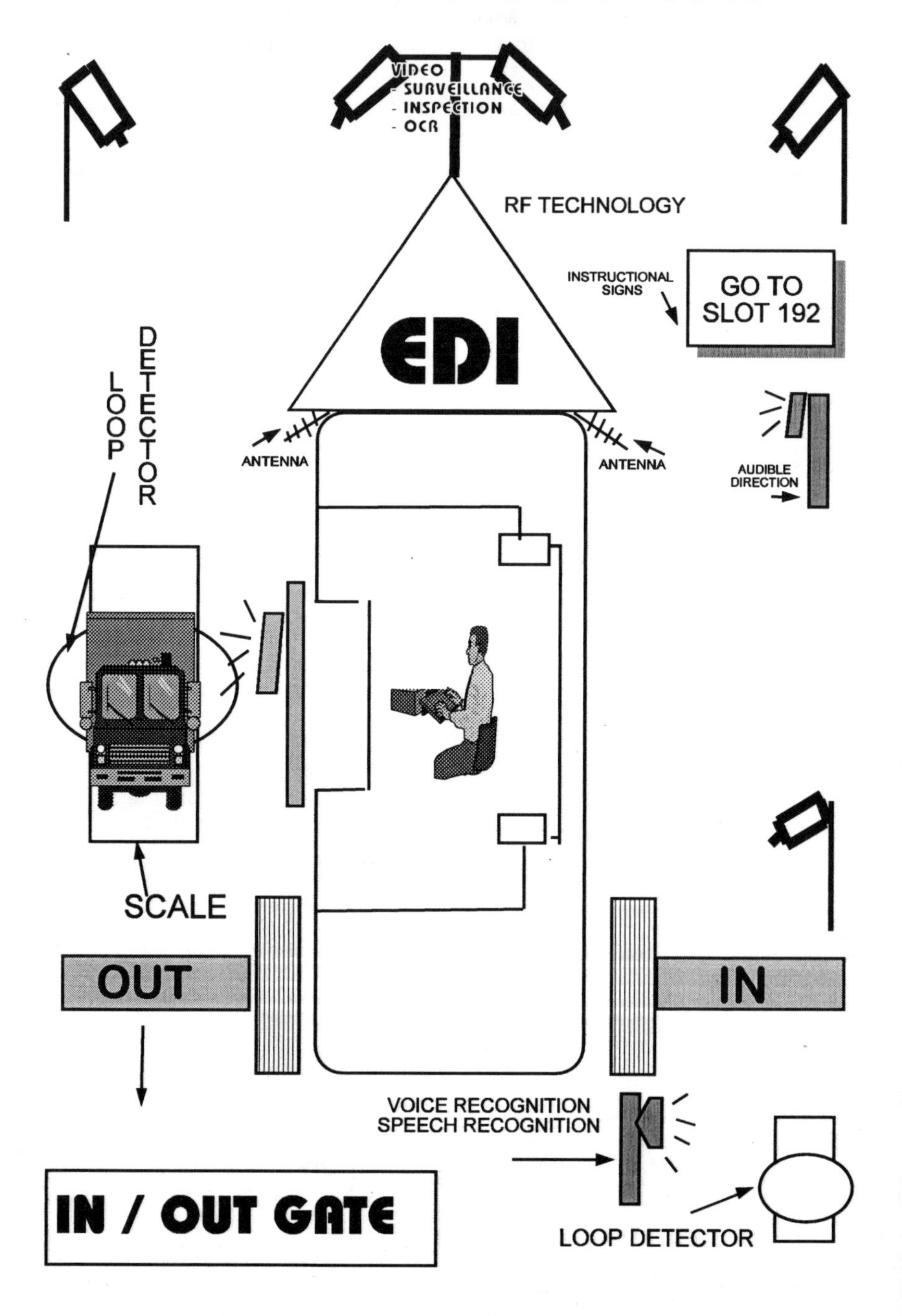

surfaced. For one thing, there is a language barrier. A driver pulls up and speaks into a squawk box and any number of languages come flowing out. Even if the gate keeper speaks the same language, there are obstacles such as noise, and garbled speech. Even when the communication is good, there are problems reading the stenciled information on the container, because the letters are blocked or covered with mud. So, at a point where it is extremely important to properly identify the equipment there are handicaps preventing a good communication. This is why RF/ID might offer significant improvements.

In these other gates the equipment that enters the yard can be identified. In the chart on page 175, there are up to three chassis- container combinations. As the equipment enters the gate area, the tractor ID is read, and some pre-gate processing can begin. Identification of the equipment could be automatically sent to the gate processing screens of existing systems, and manual systems could print the ID read from the RF readers stationed at the gate. There may be other equipment to identify and that could be done automatically as well. Hybrid systems could be employed to read the ID and register the weight.

CHASSIS GATE

The automatic identification of the equipment in our example, seven key items, will speed the gate processing. The automatic capture can provide for other applications in the process. Suppose that the owner of the chassis from the chassis pool would like to receive an automatic view of the equipment transaction. It would be possible to send an automatic message containing the chassis ID, with a date and time stamp, and location and direction to the computer of the chassis pool.

OTHER PROCESSING

Getting through the gate takes more processing. The big stumbling block is paper, bills of lading, permits, contracts, weigh slips, inspection clearances, and on and on. While most of the emphasis has been on identification there are other benefits if read-write tags could be employed. Then, not only IDs but equipment inspection records, hazmat information, bills of lading, maintenance and other information could travel with the container and other equipment.

EDI/RF/ID

For years it was thought that electronic data interchange was the answer but the reality, from my point of view, is that with read-write tags we finally have the opportunity to compliment the EDI process. Think of it as two separate systems, one (RF/ID) to emulate the paper that flows with the shipment, and the other (EDI) to follow and support the business systems. At checkpoints like the gate, the two could come together to validate the movement.

It is important to view them not as opposing views, but as two required systems that work together to insure that the transport process works. These include the EDI systems that provide shipper to consignee notifications with movement notices along the way. The RF/ID tag travels with the valuable goods within the container. Keep in mind, that EDI provides the validation or confirmation so that the right goods show up at the right location. So when the driver shows up at the door and hands over the paper work to process the order the gate attendant is likely to say, "YES, this is the shipment we have been waiting for!" RF/ID read-write tags could assist in automating the process. The two technologies, EDI and RF/ID, may someday join together in streamlining the process. This will make the data flow and the physical flow work together.

YARD

Once the container reaches the yard there is an inventory control problem. Imagine 2,000 containers spread out everywhere. Most people would find it difficult to keep track of 2,000 things. What about keeping track of 20,000 things in the same facility. This is the magnitude of the problem.

Suppose you could identify the parking areas for each one of the containers. Most yards have designated areas that are identified by lot name, and some often use row or even a slot designation. It is like a parking lot, so the particular position can be identified. There are planning systems to say where a container should be parked and there are systems that keep track of where the container is stored. So what is the problem?

Here, again, a planning system can assist in the placement of equipment for optimum stow and load plans, but there is no guarantee that the driver will place the container in its intended location. Even when the drivers try to, they might not be able to do so. When the crew takes the inventory on a soggy piece of paper, transposition errors and key entry errors still keep us from knowing where the container physically resides. Even when each step is performed satisfactorily, the container may be moved before the system can catch up. If there are 20,000 containers, and 85% accuracy, then 3,000 containers are lost! Perhaps RF/ID can assist us here.

YARD LOCATION STRATEGIES

Since the location of the container is an important part of the yard operation, how does one keep location data bases from getting 'out of sync' with reality or how to keep reality 'in sync' with the system? Most operations are twenty four hours a day. The peak activity times are when the data is needed most, and often when it is hardest to control. Thanks to advances in technology there are now three ways to approach the problem of location with the RF/ID system. 1. Wide Area 2. RF/ID Row/Slot 3. Hybrid Location being asked or required from the system. Is a very specific location needed or is a general area good enough.

WIDE AREA LOT LOCATION

The Wide Area lot Location is a unique concept. It uses a time division multiple access. TDMA, protocol RF/ID system, provides the tag IDs found in a general area in the lot. There are unique programming routines to determine yard locations.

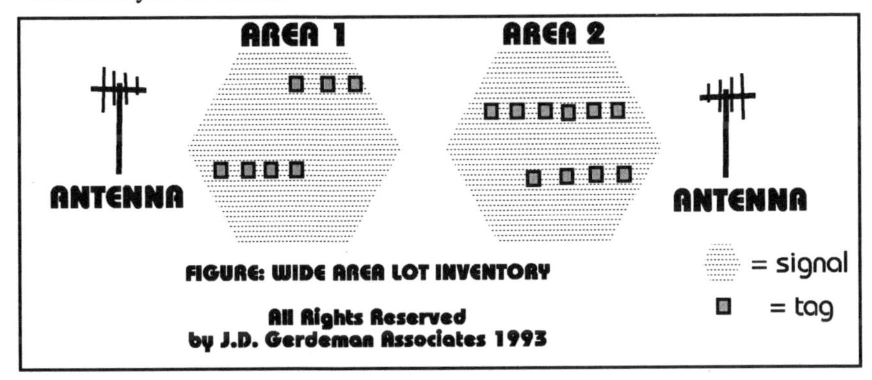

FIGURE: WIDE AREA LOT INVENTORY

In this approach antennas are located around the parking terrain at strategic locations. Special routines are used to control the antenna and to perform analysis. The idea is to locate the tag within a subset of the entire facility, or, to identify tags in a lot separate from the main terminal area.

RF/ID ROW / SLOT

Early in the application of RF tags on containers and trailers, a concept of not only placing the tags on the equipment, but that of placing the tags in the parking lot pavement was used. This concept provided a way to identify the beginning and end of a row or the end of a parking slot. If the question is, “What row is the container in?”, then the rows should have tags. If a particular slot granularity was needed, then a tag was placed at each slot. A vehicle called a mobile inventory vehicle or MIV is used used to determine the locations.

MOBILE INVENTORY VEHICLE (MIV)

The concept of MIV uses a pickup truck or hostler to locate the inventory in the yard. The truck has an RF reader mounted in such a way that the tags can be read. For example, an antenna is positioned to read tags on containers and another antenna is pointed downward to read the tags in the pavement. When a tag is read from a container, another tag in the pavement is read also and this is a designated location for the container. Multiple readers and antennas may is a designated location for the container. Multiple readers and antennas may be required. To communicate the location data to the host, a data radio is used. This concept has significant potential. A vehicle mounted with RF and communications equipment, drives up and down the rows of containers reading the container IDs from the transponders mounted on the container. Then a second antenna determines the location of the container by reading transponders located in the drive way. The determination of the read location was not necessarily exact.

Concepts like beginning and ending of rows can be used to get a general idea, but fairly exact, location of the container. To be more exact, vendors can provide pavement tags for each slot, to designate a more exact location. The exactness of the location is somewhat dependent on the route the vehicle takes. The tags tend to move within the pavement, so this shifting causes some discrepancies, and still, other tags will leave the area or become positioned so that an erroneous location can be indicated.

These things are trivial problems but none the less a concern. The tags can be mounted in the pavement such that they are less likely to move with the freeze and thaw of the seasons. Some tags are read very well when they are mounted in cement. Others need a line of site without obstruction. Some tags cannot be read through water, but can be read through snow or ice. In general, placing transponders in the pavement can be a very good way to locate a reader location.

HYBRID LOCATION

There are ways to locate the equipment when ground tags cannot be used or are not desired. This issue may become one of cost. Accuracy versus cost needs to be reviewed in light of all of the parameters. The truck is still involved with this application, but the location mechanism is different.

A. Triangulation
B. GPS
C. Combination Location

TRIANGULATION

This technique involves using radio communication signals and stationary antennas placed at fixed and permanent locations at the yard. In fact, the antenna may be at other locations, but in a relatively close location. Mathematical formulas are used to calculate the signal timings from the stationary antennas and the moving vehicle.

GPS

Global position sensing is used to determine the location of the MIV

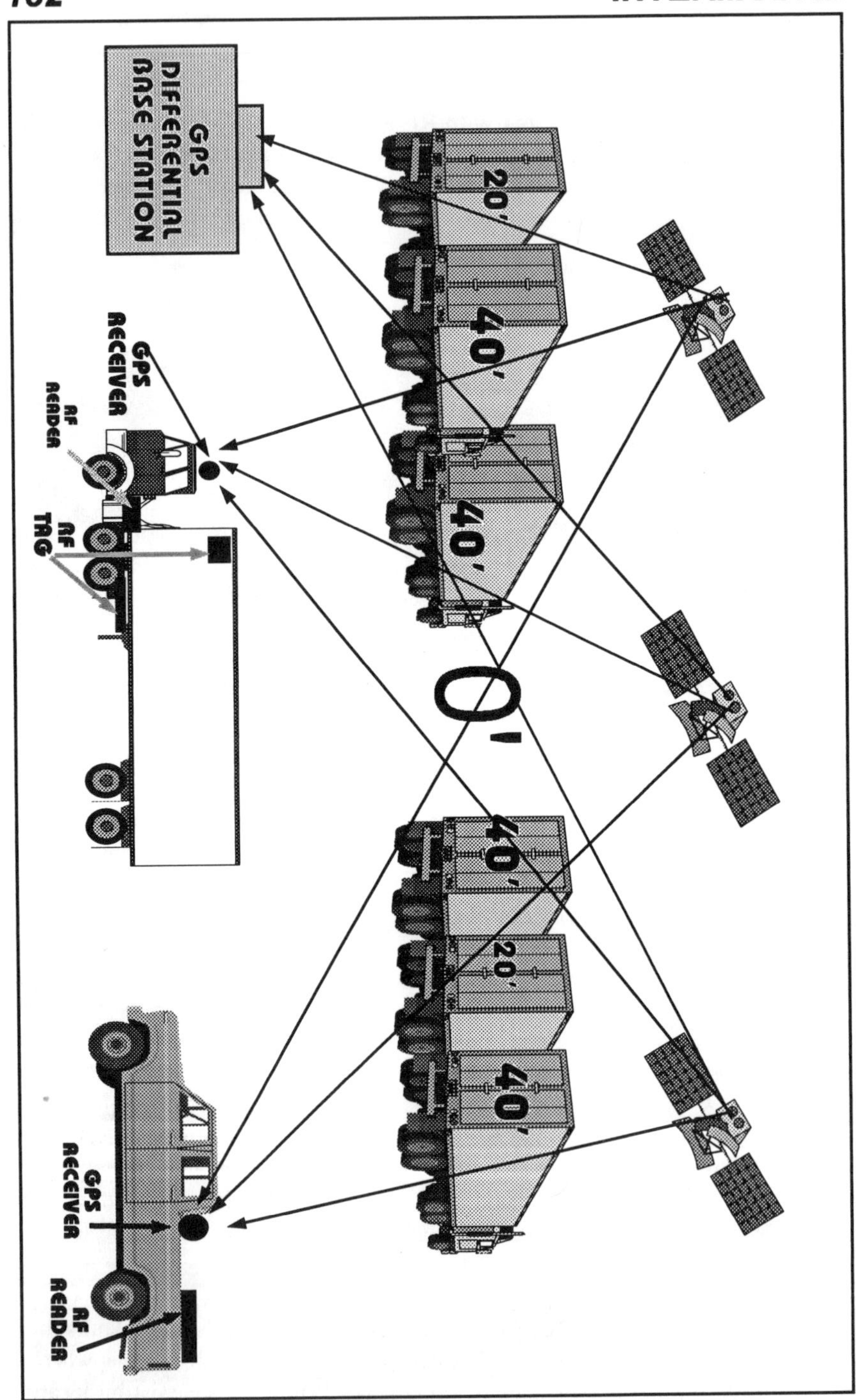
GPS
DIFFERENTIAL
BASE STATION
20'
40'
40'
0'
40'
20'
40'
GPS
RECEIVER
RF
READER
RF
TAG
GPS
RECEIVER
RF
READER

vehicle. Satellites are positioned inorbit such that multiple satellites can send location information to a special device on the vehicle. This concept is similar to the triangulation method but with RF and publicly available facilities. In addition there are routines to use a differential technique to achieve very high accuracy. The differential technique requires a permanent land station that sets he corrections required. It is estimated that the location accuracy is within inches.

COMBINATION LOCATION

Of course there are problems. Above ground tags have had some amount of difficulty. GPS seems to be a good idea until you determine that a clear line of sight is needed to the satellites. If you are under a roof or overhang or, for that matter, under a tree, communication is restricted using GPS. City caverns may block the line of site required by the satellite. Therefore, a concept that uses a combination of multiple technologies may be a better solution. As an example, the GPS system with a set of ground tags for areas where GPS cannot operate properly, or perhaps the dead reckoning systems can be used with a map matching augmentation.

LIFT

There may be some risk in using a term like lift, but it is used, and at the risk of offending those who may want an entire list of terms, I am going to settle for lift. When a container moves from one location like a spot in a yard to a rail car the container needs a lift. It needs to be picked up and moved. Special equipment like gantries, cranes, spreaders, and straddle carriers, all lift the container. These are used in the stowing, stacking, and loading processes. The technologies found inside the new equipment have interfaces to help us collect data. The more advanced the onboard processors are on these devices, the more sophisticated the operation of identifying and tracking. For those devices that are not so equipped, a retro fit operation may be the answer.

The concept is that an antenna system and a reader can be installed to read the RF/ID tag of a container and/or chassis. Then other sensing information like the position of the spreader bar, open or shut latches, height indicators, or other additional data can be captured. In fact, if the container ID, date and time stamp information along with a position or location is associated with the opening or closing of a latch, then some valuable identified location information can be recorded.

It is not always possible to get the antenna to operate properly on the equipment, or to stay in place for that matter, but when the system works there is a big advantage. This type of system can be used to determine where a container is in relation to a stack of containers. It might also assist in locating the bay and level in which the container is stored.

PHOTO BY BURNAM

PHOTO BY BURNAM

RAIL - INTERMODAL RAIL

The intermodal rail applications begin to challenge the original identification technology of the rail cars. The rail cars have two tags placed on opposite ends of the car on opposite sides. This is the case for the intermodal rail cars in North America. In addition, the track side reader system must identify all of the containers on the car. Containers can be stacked and are different sizes. The system must read the tagged containers while the train is traveling at high speed. Wheeled operations have the containers, the chassis, and the rail car to identify. When too many chassis are sent to one end or the other of the system, they may need be returned to another location. Multiple stacked chassis can be added to the rail car, and the reader system is required to read it. The identification system needed to read all of these combinations of tags will require addition antenna and reader capability, than that of the rail only configuration.

MANAGEMENT

Management systems are required to get the most out of the identification system. The automatic nature of these systems provide a significant opportunity for advancements in customer service and operational improvement. Real-time status reports provide management with real-time data. 'What if?' questions will tend to be more accurate. With accurate location data and equipment identification, detail load planning systems and stow planning can be more accurate and effective.

SUMMARY

Intermodal application of RF/ID is driven by the fact that so many players must share in the movement of a shipment from beginning to end. Efficiency in determining the identification of the equipment used in the process, will improve gate processing and yard inventory. New systems provide a greater opportunity to be competitive and to increase customer service .

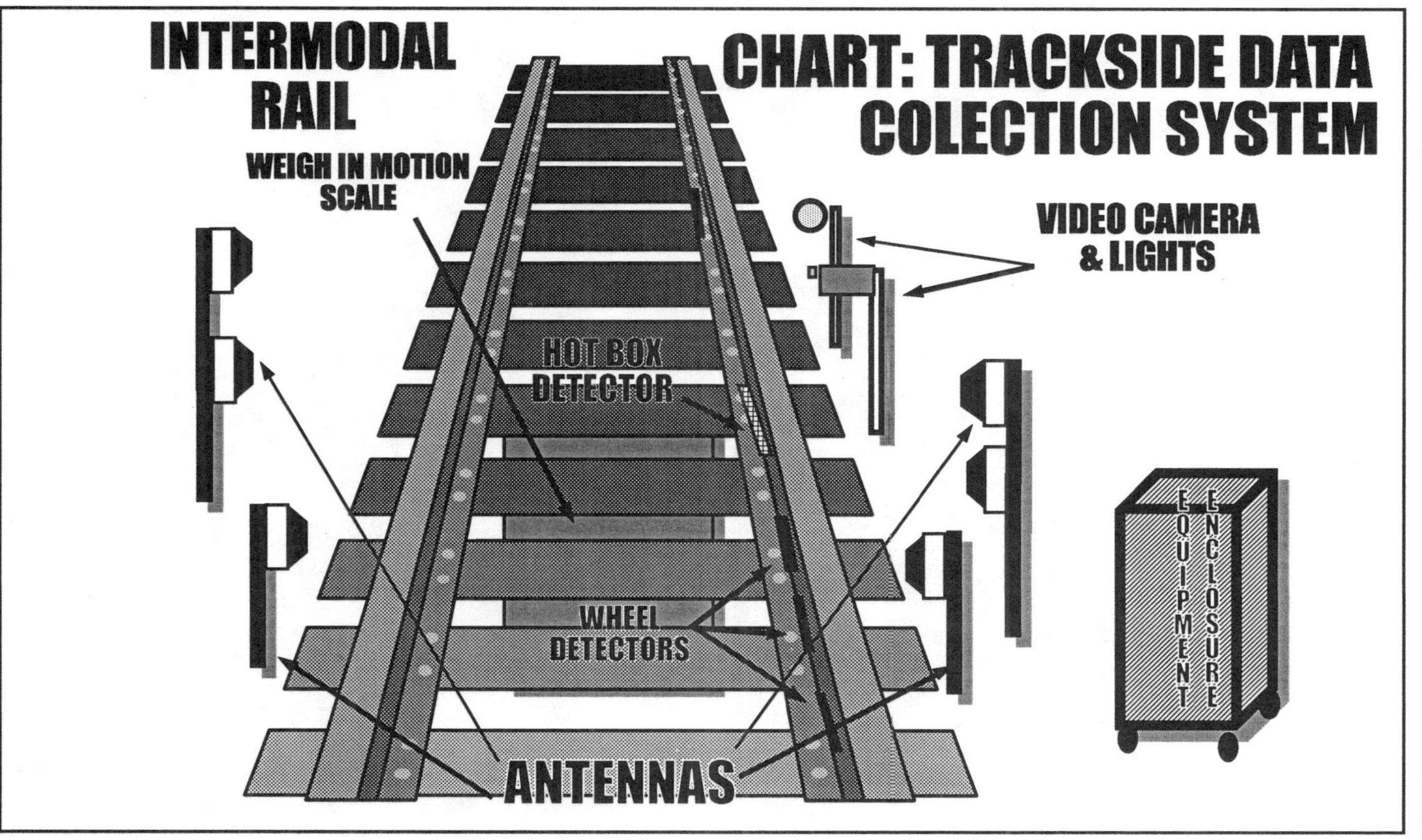
INTERMODAL
RAIL
CHART: TRACKSIDE DATA
COLECTION SYSTEM
WEIGH IN MOTION
SCALE
VIDEO CAMERA
& LIGHTS
HOT BOX
DETECTOR
WHEEL
DETECTORS
EQUIPMENT
ENCLOSURE
ANTENNAS

INTERMODAL RAIL

ANTENNAS AT MULTIPLE HEIGHTS

PHOTOS BY BURNAM

CHAPTER XVI

FLEET MANAGEMENT

Fleet managers have significant problems for which many of the computer systems architects have solutions. The collection of sufficient data often gets in the way of progress. RF/ID offers a way to collect data automatically in the course of conducting the business at hand. Fleet management is getting more complicated all the time. Road taxes and permits, licenses, insurance, safety, administration, maintenance, the list seems endless. Fortunately, computer systems help us get through the maze. There are fleets everywhere, buses, trucks, vans, car pools, rail, taxis, limousines, school buses and rental car fleets and more. It is one thing to know the particular fleet, and even the specific problems associated with each, but it is another thing to impact the operation of the fleet in such a way that you can make a significant difference to the bottom line.

COST DISTRIBUTION

Each fleet has its own cost structure. Drivers, fuel, and maintenance are all costs that have significant impacts from an RF/ID point of view. The Cost Distribution Chart on page 190 shows a relative comparison of the costs. Drivers, fuel and maintenance is about 70 % of the equation according to the American Trucking Association. That leaves the remaining 30 % for taxes, insurance, administration, ownership etc. Your own fleet may have other ratios, but these are the industry averages.

Productivity and cost reductions are always a concern of the fleet managers. The use of knowledgeable fleet managers is a successful tool for the owners. Turnovers and shortages of skilled people have caused some less than optimal operations from time to time. Fleet sizes have grown, and maintenance staffs have been reduced or stretched to their limits. In general, there is not a glut of resources to tackle the problems. All of this leads to the question of what to do next.

DATA TOOLS

Fleet managers need more data to analyze to make informed decisions about the actions needed. There are software packages available that will analyze just about anything for fleet managers. If something new is needed, the popular personal computer or laptop has capabilities to assist in getting the data needed.

The problem with data is that it often costs too much, so that it inhibits solution development. The manual processes to capture and enter the data into the system can be very costly. What is needed is the ability to capture the data automatically during the normal course of business. Now we have the ability to automatically capture data. Both radio frequency identification systems and on

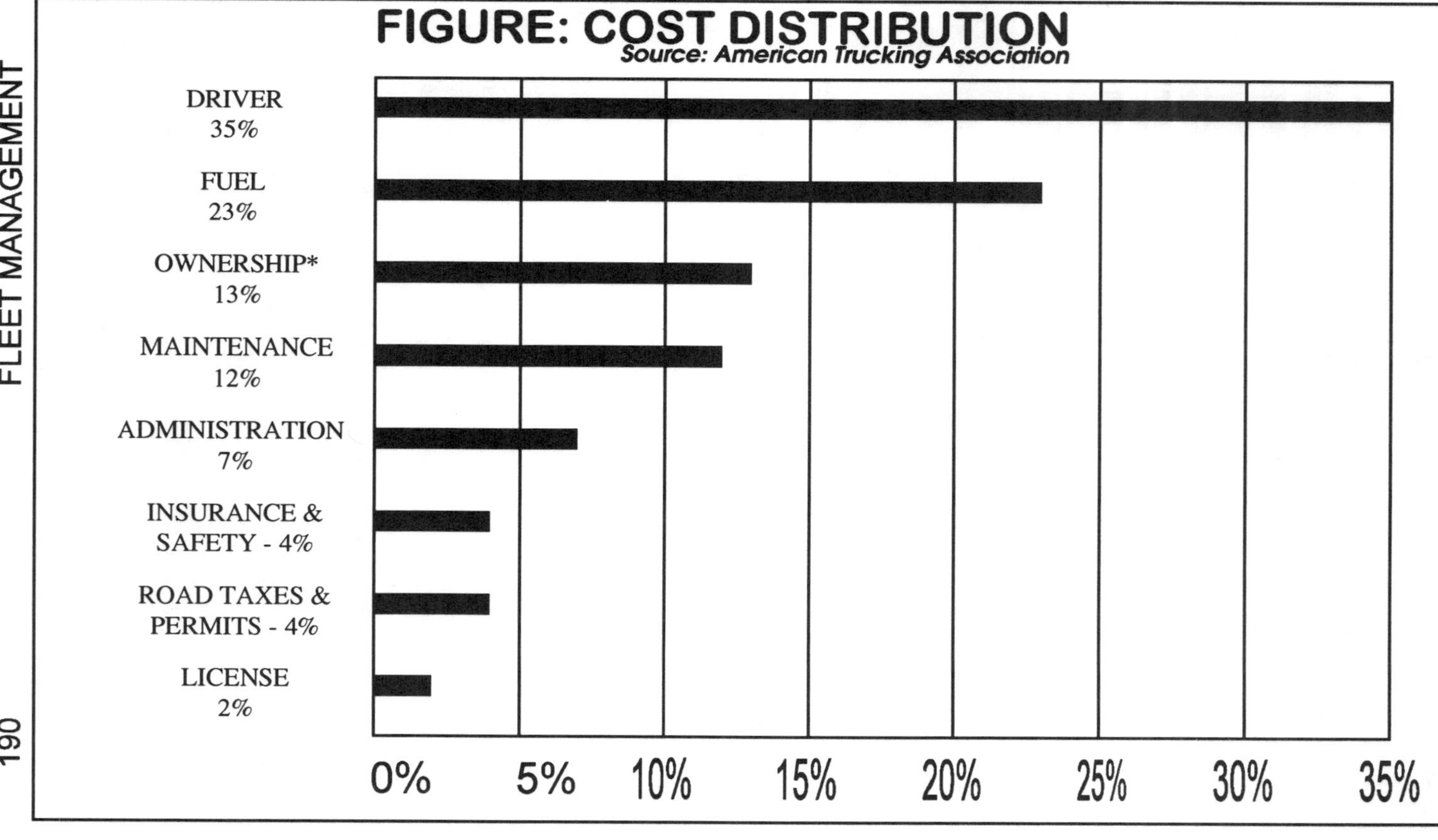

FIGURE: COST DISTRIBUTION
Source: American Trucking Association

TOP & BOTTOM PHOTOS BY BURNAM - CENTER PHOTO BY GERDEMAN

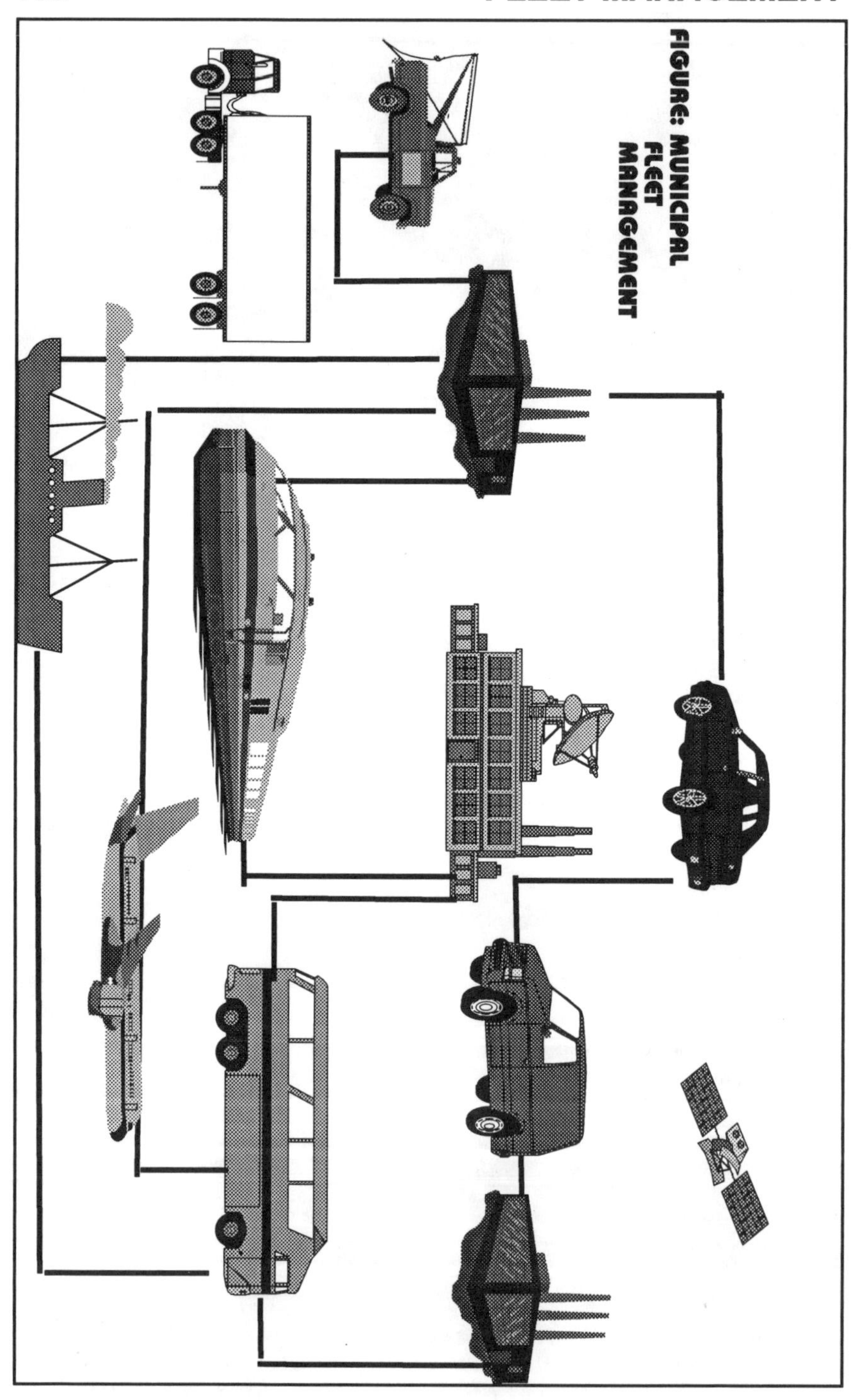

FIGURE: MUNICIPAL FLEET MANAGEMENT

board computers and personal communicators can provide data capture during the normal course of doing business. Do we need both?

COMPUTERS AND RF/ID's

Both are computers of sorts, and they serve different functions. Low cost tags can assist in the positive identification of assets. A tag can be read automatically without the driver hitting an enter key. Consider the tag as having the function of a card in the computer that can identify the asset. Both devices can provide fuel gauge read outs. The tag can be read at fixed points and the computer can send the fuel level continuously if needed. It depends on customer wants and needs. What is the requirement? The answer may be both devices. This is true when functions like electronic license plates and toll tags enter the picture. The technology should always be viewed in terms of the overall benefits.

The tag is a small device with a relatively small cost. The computer is a general device that can do everything. Both devices vie for funding. At one end is a low cost and low function item, and at the other end a relatively high cost and high function item. Adding devices and function, as in the dynamic tag, and the read-write tag, the tag system starts to look more like a computer. Then the system is doing more. The computer has general purpose communication functions to draw on and can provide every bit of data supplied by the tag system. A computer is usually in control of the tag reader, so again, both are needed.

BUSINESS OBJECTIVES

Fundamentally, the objective may be to earn a profit. In the public sector the objective may be to stay within budget, or to reduce the budget. Whatever the overall target is, there are factors that contribute to funding the operation. These factors include improved productivity, cost reductions and customer satisfaction. There are a number of other factors that, either from a departmental view or departmental focus, will assist in meeting or exceeding the overall objective of the organization.

If the objective is to extend the useful life of the assets, then it is beneficial to keep the assets in service for as long as possible. This will delay the need for new capital to replace the equipment. Radio frequency identification can assist us in this effort. If the read-only tags are used, then the reliable identification of the asset is possible at strategic points like in/out gates and maintenance bays. This would guarantee that actions taken are recorded in an accurate way. Other uses for the read-only tag might include automatic operation for refueling vehicles, and the automatic start of washing equipment and the like.

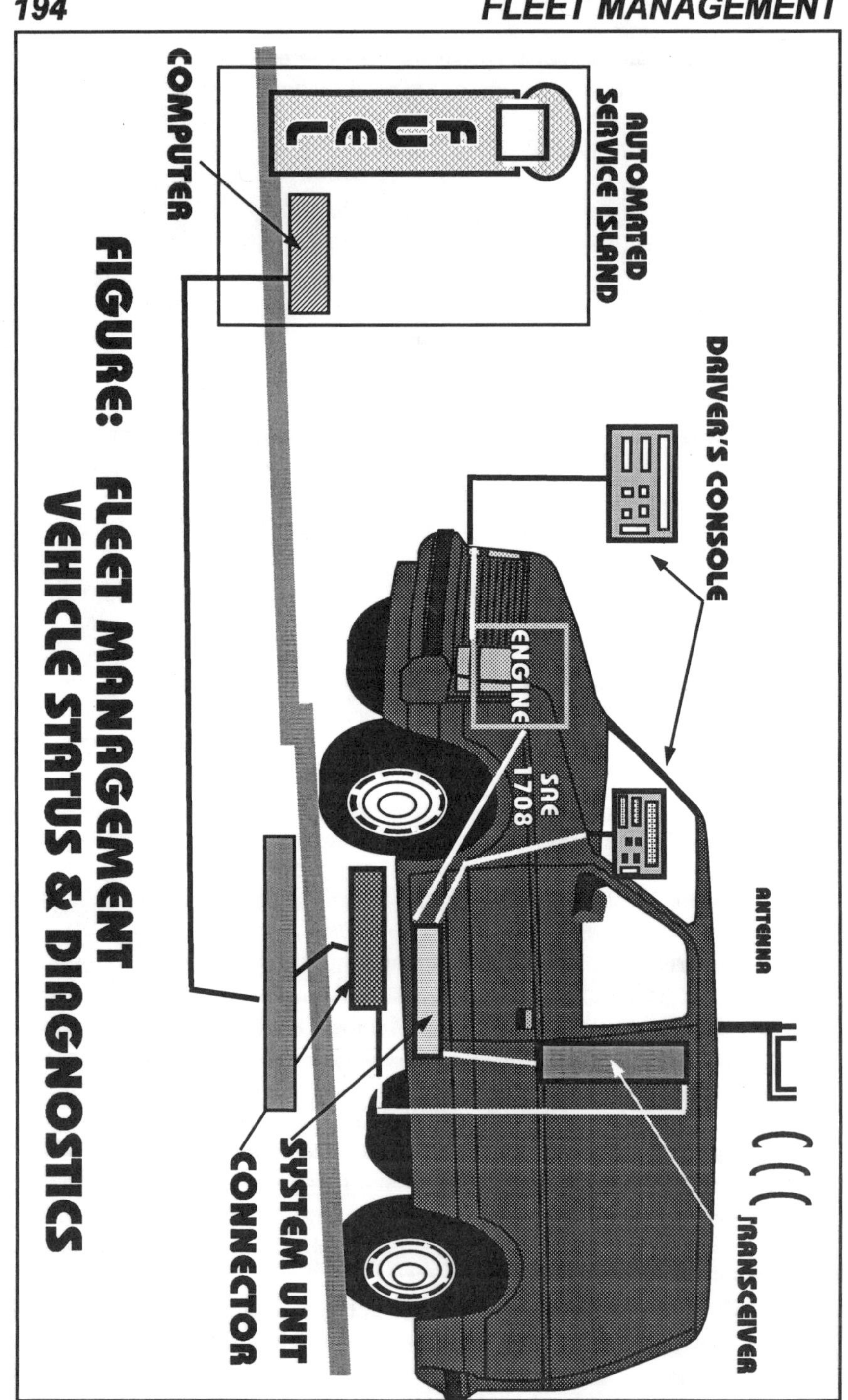

FIGURE: FLEET MANAGEMENT VEHICLE STATUS & DIAGNOSTICS

DYNAMIC TAG DATA COLLECTION CONCEPT

However, if the dynamic tag were used, then it would be possible to read the ID, fuel level, mileage, and other information like oil condition, etc. Now it is possible to collect the data elements automatically in a timely fashion during the normal course of business to make decisions closer to real time. A trigger for the time to change oil could be based on miles traveled since the last oil change, or it could be based on the condition of the oil in the vehicle. This could assist in extending the life of the vehicle and potentially save money used on oil.

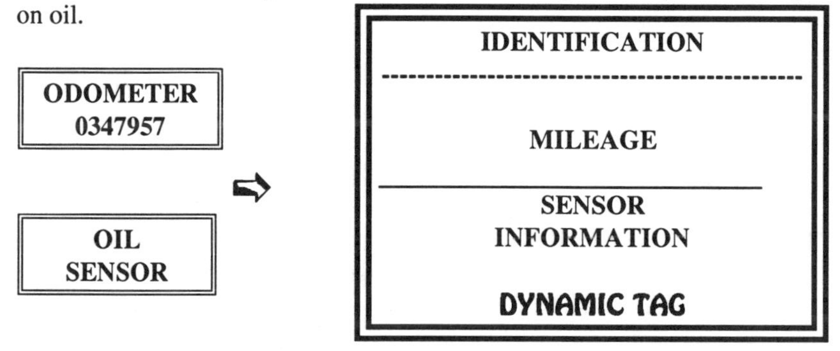

FIGURE: EXAMPLE OF DYNAMIC TAG COLLECTION

The output of the dynamic tag increases potential improvements to costs associated with maintenance and fuel. The potential for savings is generated by the use of RF identification tags and the dynamic tag concept for all of the key pieces of equipment like tractors, trailers, trucks, wheels, and engines.

The chart on page 194 shows how sophisticated computer collection and transponders might coexist. The application shows a refueling station. A transponder mounted on a vehicle interacts with an antenna in the surface of the roadway, and this interacts with a computer in the automated service island. While the concept can have many different configuration options it is important to further explain this particular configuration.

The system unit of the vehicle is connected to a driver's console. The console can be mounted inside the cab or could be removed for use away from the vehicle. The diagram shows a physical line connection, but the communication from the console to the system unit might be a physical connection, a short range data radio connection or another options. The driver's console may have sufficient capability to act as a stand alone device to capture data when the driver removes the console from the vehicle. Later, the driver could make a physical connection to the system unit and data could be sent across a data radio connection using cellular, radio or satellite communications.

In the concept shown, the system unit could be used to accept data from the

J 1708 Bus and used to monitor the engine health and status. Certain data can be collected and analyzed for later transmission to a host through the communications capability mentioned above. The processor or system unit could also fill in a data field of the dynamic transceiver. In this way, the system is configured for a flexible approach to data collection.

READ / WRITE POSSIBILITIES

From the viewpoint of fleet management, there are a number of read / write applications that are possible. Maintenance records and inspection records can be written to the tag, to be read later at other points along the route. When the fleet is managed with multiple maintenance centers, some of which are contract centers, the information could travel with the vehicle. The hazardous material records could also travel with the vehicle.

Transponders contained within a tire could hold the ID and certain maintenance functions like location pressure and service dates. There are even ways to indicate current pressure.

The driver ID could be written to the vehicle tag. As the vehicle travels along its route, the records from strategically placed readers could perform additional data collection. This might then provide data base analysis of performance and productivity. If the driver were to have his own RF/ID tag, the data collection system could read the tag, and the information on the trip, schedule, and route could be written to one of the transponders. Again, the data is captured through the course of performing the work, and needs little intervention. Where there is a need to capture the contents of a trailer or truck and produce contents paper work like bills of lading, then this information could be written to the transponder.

PUBLIC SECTOR FLEETS - OTHER MULTIPLE FLEETS

When you consider applying RF/ID to fleets, there may be advantages such as relating multiple fleets within the same system and taking advantage of the activities within one system to help another. Public sector fleets could use the IVHS application to tag vehicles like buses or police cars and track their progress through the streets to determine traffic flow. Other application linkages are possible.

Servicing vehicles could be scheduled based on data collection from security vehicles, as an example. Dynamic tags on garbage containers indicate a load level. A safety vehicle routinely traveling through the neighborhood reads the level and location. Later when the vehicle returns to the post the data is sent to the garbage pick up service. A route planning system sets a route to pick up the load. The system could save fuel and driver time and still maintain a high service rate. One fleet helps the other. The data is collected in the normal course of doing business.

SUMMARY

Fleet managers can obtain information for decision making automatically in the normal course of business operation using radio frequency identification. RF/ID and computer systems can coexist.

CHAPTER XVII

JUSTIFICATION

Using radio frequency identification is justified by the improved timeliness and accuracy of data collection. With the right information at the right time, decisions can be made that affect the bottom line. The system is more productive. Justification means to have a good reason or excuse. This chapter explores the reasons for the use of RF/ID in operational systems. The investment needed to implement the radio frequency ID is not trivial. There had better be significant reasons to justify the expenditures. One flag ship company has committed $35 million to tag their containers. The North American Railroads will invest over $300 million for AEI. The European Rails will invest more than that for their systems. Trucking companies are spending significant dollars to tag their fleets. There must be a good reason for all of this investment.

VALUES

There must be a value associated with the specific application of the radio frequency identification system. The value that the application brings to the organization may manifest itself in many different ways. Often we speak in terms of tangible and intangible benefits.

A.	Time Savings	B.	Safety
C.	Security	D.	Control
E.	Performance	F.	Convenience
G.	Cost Savings	H.	Entertainment
I.	Information	J.	Efficiency
K.	Productivity		

All of these things brings to mind particular images. Time savings can mean labor cost reductions and increased work production. Safety is often a reason for spending resources and also for the entertainment value they perceive to be available. Some want information because knowledge is money. The value placed on the situation, product or system will help evaluate the product or service. This provides the reason or excuse for the investment. Management teams would prefer to have the facts surrounding the decision. They would like the rationale for making the decision. They want the justification, the excuse, the reason why this is such a good idea. They want more than the intuitive knowledge that this is the best idea around.

The justification that is provided to management should reflect the values they perceive as important, as well as, be in terms that everyone can understand. So there needs to be a dollars and cents strategy. There must be a justified explanation of cost and a description of the tangible benefits.

MYTHS

We all want savings including the soft dollars. Well, what is a soft dollar? Most money earned is the result of hard work. For example, if 20 per cent of the work force is eliminated the justification is 20 per cent. However, if the operational management will not eliminate this 20 per cent, that is take a 20 per cent cut, then there is no justification. My suggestion is to define the justification in terms of hard dollars. If the manager can commit to the 5 per cent then use it. Real dollar savings are best.

Sometimes the dollar savings that justify a given application are used by other changes in the organization. If five projects claim 20 per cent improvements for the department, then at the end of the implementation of all five projects, the department should be eliminated (5 x 20 = 100). Well, not really, but there should be a very significant decrease. (100 - 20 per cent x 100 = 80; 80 - 20 per cent of 80 = 64...) Something like that is possible.

One common trap is that multiple projects claim the same dollars as justification when the viewpoint should be from one point or the other. Be certain that you are looking at the savings and costs from the right perspective. If another change is claiming justification and the change has not yet been implemented, it might be difficult to know if the justification is valid.

The viewpoint and the justification of any major investment should not come only from one department, but from the total organization. If, in an attempt to save money, the law department cuts its staff from 30 people to 20, and the departments who were served by the 10 eliminated lawyers, now engage 15 lawyers to handle their work, there is no savings. Instead, there is a cost increase.

There are applications where departmental improvements of 40 to 50 per cent are possible, but keep in mind that every idea does not have the same benefit or value. Check the estimate against reality. Determine the impact in other terms like people, or dollars. Is it possible to reduce the head count from 20 to 2? Is it possible to reduce the budget from 1.5 million to 0.3 million? The answer, of course, is maybe yes or maybe no. The numbers change with each situation. Inspect them from all angles.

TECHNIQUES

There are many proven techniques for gathering data to answer the justification questions for a changing process. A survey is often effective as a simulation of the process. The point is that there should be a technique that is suitable to the organization. The process for gathering data to justify the project must be scrutinized as much as the process being changed. The technique must gain the confidence of the organization for it to be effective. Thinking through the participant list, the reviewers and the study team, well in advance of performing the justification study, is time well spent. There needs to bc credibility associated with the project.

The survey techniques used may vary from project to project and organization to organization. There is not just one way to approach the justification process. In fact, as new concepts and approaches are being discussed for the changes needed in the business, new approaches my be needed for the justification. This is an area for changing the paradigms. One way of approaching the problem is to use a technique known as a Joint Application Design.

The Joint Application Design technique is set up with a high ranking management sponsorship and high level management involvement. The technique is flexible in approach but very powerful in producing results. There is a session leader and participants assigned. A recording secretary is also part of the group but has been designated for the job as opposed to being chosen from within the group. The participants can have varied responsibilities but should represent a good cross section of the organization as well as multiple opinions within the same disciplines within the organization. The concept is to re-engineer the processes that can have significant overall benefit. The sessions usually begin with a description of the current processes followed by a discussion about the desired changes and a description of the new scenarios.

Pick the technique that best suits your organization. Executive sessions can pinpoint the strategic problems that must be addressed. Then, implementation sessions, such as the joint application design, explore the details required for implementing change.

JUSTIFICATION PROCESS

The goal of this process is to provide an evaluation of the costs versus the benefits of the proposed changes. What is the implementation of the radio frequency identification system worth to the company? What benefits will the company realize? The process includes:

1. Understand the current operational scenarios.
2. Determine alternative sets of scenarios.
3. Identify departmental and operational benefits.
4. Review recommended scenarios with the key players.
5. Make the decision for next actions.

There may be a range of benefits for each change. It is imperative that the reasons surrounding the delta are well understood. These reasons should be part of the supporting documentation. The specific organizational criteria should be part of the descriptions used. In any justification there needs to be a clear understanding of the objectives. There are considerations given relative to the approach used on such items as costs, capital, and expansion strategies. There is also a need to understand the value of the affected assets and other plants and equipment. All of these factors need be considered within the strategic objectives of the organization. There are three fundamental considerations in this process that are the focus either individually or in combination.

1. Cost Reduction
2. Productivity Improvements
3. Revenue Generation

COST REDUCTION

At the core of cost reduction strategies is the notion of maintaining the same volumes, but gaining a competitive edge by performing the function at a lower cost than the competition. Clearly, if current volumes could be maintained, and the cost of doing business could be reduced, then an increase in profit is possible. The goal therefore, is to reduce the cost of doing business. This approach is a good one, but does not often have a significant impact unless a revolutionary approach can be used. To get a revolutionary approach, a new technology must be invented to make the dramatic change possible.

PRODUCTIVITY IMPROVEMENTS

Productivity improvements, in this context, means getting more from the capital that is invested in the business. There are rather large investments in plants and equipment. Depending on your employment practices, there may be a large investment in employees as well. So the objective is to get more work from the same resources. This does not mean longer hours, but rather a more productive work day; increased load factors, increased loads per hour, decreased gate processing times, and decreased lifts per hour.

REVENUE GENERATION

Increasing the amount of revenue generated can also have a significant impact on the organization. The organization may gain in the market share with a particular service offering or service level. A unique offering will drive customers away from the competition.

In the three areas of cost reduction, productivity improvements, and revenue generation there is a need for significant change. To effect significant change, a significant breakthrough in doing business is required. These breakthroughs often require a revolutionary technology applied in new and creative ways. The radio frequency identification systems have that kind of potential. New processes can be implemented and provide that competitive edge.

JUSTIFICATION BENEFITS

Value considerations are a specific value placed on a specific application. The savings might include time or resources, or increased productivity.

Increasing the flow through a toll road, without extra facilities, is a significant benefit. The price or cost associated with the change influences the decision. The world is driven by costs. So costs must be the most important part of the equation.

The justification for using a radio frequency identification system begins with the benefits of the accuracy and timeliness of the data capture process. It is the ability to collect the ID automatically, and in the course of normal business processes. The nice thing about RF/ID is that more than the ID can be made available in this same process. So there is a potential for more benefits than a simple identification. Improving resource utilization through improved data accuracy and/or data entry timeliness, results in operational improvements. The benefits come from three areas:

1. Labor savings
2. Resources/Asset Utilization
3. Customer Service

The elimination of a percentage of the labor force may not be necessary for the benefit to be seen. If a gate attendant can be more productive, more drivers may be processed in and out of the gate. In this case, more productive gate attendants is enough justification. In other cases there may be a direct labor savings. As an example, if 6 gate attendants can accomplish the same work as 8 a savings of 2 might be possible.

Utilization is another primary area for justification. In this area consider the additional use of the resources, or better control and thus reduced loss or loss of the use of the resource. This justification area includes insuring processes like maintenance are performed, increasing the useful life of the equipment. Providing an accurate inventory of the equipment assists in the better utilization of the equipment.

Customer service justification offers a wide range of benefits. Knowing where the equipment is located and having the ability to quickly answer customer inquiries about shipment status is of vital importance to shippers and carriers.

IMPROVEMENTS

There is a need to understand where possible improvements can make a difference. It is also important to understand what must be done, what is done, and what could be done with the resource. If there is unimportant work being performed that does not add value to the process or the company, or is needless effort, then it is not required. It is important to reflect on what the proposed changes can do to provide improvements. Artificial benefits will result in border line benefits. Here are some specific improvements to look for when you are processing the justification.

A. **Direct People Savings**: If the savings are determined to be 15%, and there are 100 people doing the work, then 15 people can be eliminated from this responsibility. Of course the catch may be that the 100 people work in ten different locations and that even with the new process, nine people are required at any given location. So the possible savings are 10 people rather than 15. (100-(10x 9)) The point is that each location must be reviewed for specific requirements. Peak loads and the distribution of labor should be a concern.

B. **On Demand Labor**: While there are labor laws and specific contracts to be considered, there are times when it might be possible to find potential savings from the early release of the work force. For example, look at a situation where, at the end of a shift, a crew of 24 workers are used to document the location of the assets. This takes an added 15 minutes. If the application of RF/ID could eliminate this step then 24 workers could be released 15 minutes earlier every day.

C. **Reduced Search Time**: Determine what is normally required to search and find an asset that now can be located through the controls and tools associated with RF/ID. There is more effort expended here than we admit. If this effort can be reduced there is a direct labor savings.

D. **Rework Labor**: Sometimes work is done twice. Mistakes are made because the goods or the container in which the goods were shipped were not properly identified. When the error is found the items need to be returned, and paperwork to redirect the shipment and the actual work to get the shipment to the correct destination is required. With RF/ID and the associated systems these errors can be eliminated.

E. **Extending Equipment Life**: If the RF/ID system makes it possible to collect maintenance information automatically, the life of the equipment can be extended. Repair alerts can be automatically sent to the maintenance and repair shops and actions can be taken based on informed decisions thus providing the savings.

F. **Equipment Utilization**: On a system wide basis, planners can see if equipment is migrating to one end of the system or another. When equipment leaves the maintenance bay, it can be added automatically to the available queue.

G. **Customer Service**: Customer service options are now available with the RF/ID system that have never before been available. A notice can be sent to the customer from automatically reading the transponder. Special services for processing through the system can be offered with the use of the tag technology.

COSTS

The cost of making the change needs to be determined with the same vigor as the savings potential. The cost of new technology involves a number of factors.

A. Cost of Data Collection Equipment
B. Cost to Install the Equipment.
C. Systems Development Costs
D. Maintenance

E. Training Initial and Ongoing

F. Transition Costs

G. Construction

These are some of the major categories of costs associated with implementing the new system. While most of the costs are obvious from the point of view of new computer systems, the construction cost is often overlooked. Antenna placement, and other control equipment like loop detectors, are cause for significant construction and civil engineering work is required. Even when placement is provided on existing structures, there is a need to run power and other cables to the equipment.

One thing to keep in mind is the cost of the transponders. An initial expense for the RF/ID system is the cost associated with placing the tags on the equipment. This cost is often found to be less than 1% of the cost of the equipment. Once the tags are placed on the equipment, new applications do not need to absorb the tag costs. So if the tag is placed on the tractor for a toll road application it can be read at the company's gates. This will provide additional opportunity for new applications.

SUMMARY

RF/ID provides significant justification opportunity for changing the operational processes. Cost savings can be found in labor, utilization, and customer service. Specific processes discussed throughout the application found in this book can be examined for justification. Design and functional capabilities can bring out justifications from revolutionized processes. Imagine the possibilities.

CHAPTER XVIII

VENDORS

RF/ID has just begun to make an impact on our world. Many manufacturers of RF/ID equipment have tried their luck at this and some have been winning and others have lost the battle. There are new, powerful players entering the market that offer significant products to fill the user's strategic application needs. Custom solutions are available.

This chapter is dedicated to the vendors of RF/ID technology. These are the manufacturers and developers of the technology. Several years ago manufacturers would basically make to order just about anything you wanted. Some still will primarily because the business has not sufficiently matured. Progress has been made with the standards and competition is now the name of the game. As a student of the RF/ID industry, it is interesting to see the developments. The functions have increased to meet the needs of the customer. Distance, speed, and memory are three that come to mind. The ability to read one or many tags is an important concept as well as read-only and read-write systems. One of the important factors is the players in the game. Major corporations are interested in the revenue streams that RF/ID has to offer.

Some of the vendors listed I have known as a part of the application development efforts over the past six years. The information about each is a composite taken from readily available sources, like brochures and financial statements. This is not a full and complete description, but a brief look at who they are and what they offer. Here is the list of vendors.

1. Advance Systems Group International
2. Allen-Bradley
3. Amtech
4. AT/Comm Inc.
5. AT&T
6. Checkpoint
7. Delco Electronics /Hughes Electronics
8. ID Technologies
9. Indala Corp.
10. IntelliTag Products
11. Micro Design A/S
12. NDC Automation, INC.
13. NEDAP GIS
14. QED Technology
15. SAVI Technology
16. Sensormatic Electronics Corp.
17. SSI Custom Data Cards
18. Telsor Corp.
19. Texas Instruments Inc.

20. Vapor - Mark IV
21. X-Cyte Inc.

ADVANCED SYSTEMS GROUP INTERNATIONAL

Advanced Systems Group International (ASGI) is a federally designated small business corporation that offers a set of products including RF/ID. They have applications for asset management, personnel, and vehicle access control. They emphasize a number of industries including petrochemicals, heavy construction, manufacturing, transportation, and government agencies. AGSI has developed a proprietary Intelligent Tagging System™. They support a number of frequencies with their system. the emphasis at ASGI is on true integration for their products and customer information management requirements.

Advance Systems Group International
2214 Rock Hill Road, Suite 505
Herndon, VA. 22070

ALLEN- BRADLEY CO.

Allen-Bradley is a subsidiary of Rockwell International and has been providing customers with products to solve industrial automation problems since 1907. They use a worldwide network of sales and service locations in addition to a network of distributors and system integrators. They produce a number of tags for location and bin, pallet and equipment identification with read distances of six inches to five feet. The technology is a read-write variety.

Allen-Bradley Co.
1201 S. Second St.
Milwaukee, WI 53204

AMTECH

Amtech has a world wide distribution of RF identification for the transportation industry. Amtech's success is based on the identification of vehicles and equipment with a combination of capabilities they list as speed, range, accuracy, reliability, and frequency agility. Multimodal standards based on or compatible with the Amtech solution have been established by the AAR, ATA, ANSI, and IATA. Amtech provides a full support capability with R&D, Manufacturing, Computer Systems and Software, System Design and Installation, Maintenance Services, and Training, using a distribution network.

Amtech Corporation
17304 Preston Road, E100
Dallas, Texas 75252

AT/COMM

AT/Comm builds read-write systems for the electronic toll collection systems. They have emphasized their toll collection application.

AT/Comm Incorporated
America's Cup Building
Little Harbor
Marblehead, MA 01945

AT&T

AT&T's intelligent vehicle highway systems are pointed at the transportation marketplace offering commercial vehicle services, traveler information services, mobile communications and information systems, and a variety of hardware and software offerings. The AT&T Smart Card technology functions as a secure system peripheral, physically designed for multiple applications with their long life usage requirements. It can withstand ordinary credit card handling, as well as, the heat and stress of ID laminating systems. The Smart Card uses inductive power transfer, and is powered from the terminal to the card. The card has 3K of fully erasable memory and is used in toll applications.

AT&T IVHS Communications Systems
55 Corporate Drive
Building 15B
Bridgewater, NJ 08807

CHECKPOINT

Checkpoint has been very active in the retail product security and the access control application areas. They produce extremely low function, low price tags, 500 million annually. The tags are one bit tags and there is a 64 bit proximity tag available. The tag is very small, making it easily affixed to labels and boxes for control purposes.

Checkpoint
1-800-257-5540
Thorofare, NJ

DELCO ELECTRONICS / HUGHES ELECTRONICS

Delco Electronics manufactures RF/ID systems. Their products include tags, readers and associated components to meet industrial application requirements. Hughes Identification Devices apply their systems to automated material handling, automatic tool ID, delivery systems, access control and automotive ID. Hughes has been fostering an open architecture approach. Their emphasis has been in automotive, military, and security. They are active in the transportation services area as well. Their channel of distribution include manufacturers, OEM's, Systems Integrator's, and VARS.

Delco Electronics
700 E. Firmin Street
Kokomo, Indiana 46904 - 9005

ID TECHNOLOGIES

ID Technologies manufactures low frequency RF/ID products. They specialize in industrial automation systems. The low frequency RF/ID system is a passive system. They produce tags, readers, and system software for identification of objects and people in the industrial automation environment.

ID Technologies
P. O. Box 1437
Breckenridge, CO 80424

INDALA CORP.

Indala Corp. offers a high performance solution using RF/ID. They apply RF/ID to identifying, positioning or tracking almost anything. Their tags and readers work with reliability in environments where bar code performance is marginal. Indala readers can provide plug compatible emulation of bar code wands and lasers, as well as other industry standard outputs. Their RF/ID solutions have been available since 1985. They produce both passive and active tags with a variety of form factors available. Their installation base includes trucks, rail cars, containers, pallets, hazardous material, chemical containers, and high value products. They welcome inquiries. They have installations worldwide. Indala uses a VAR program for application solutions.

Indala Corporation
711 Charcot Ave.
San Jose, CA 95131

INTELLITAG

IntelliTag offers products in the Electronic Toll and Traffic Management, ETTM, segment directed at transportation needs. Amtech and Motorola have joined forces to bring a new vehicle-roadside communication technology, VRC, to provide traffic management flexible performance. Amtech and Motorola merge the best of RF identification and mobile data communication into a single system, capable of both lane specific and wide area communications. An open architecture adapts to a variety of interfaces and protocols, and the IntelliTag 2000 line of RF products offers the building blocks for Vehicle to Roadside Communications, Commercial Vehicle Operations and Advanced

Traffic Management System applications.

IntelliTag
8201 E. McDowell Road
Scottsdale, AZ 85252

MICRO DESIGN A/S

Micro Design A/S, located in Tronelag, Norway, was established in 1984. Their business concept is advanced R&D of electronic products, hardware and software. They have strong resources, combining MICRO electronics, MICRO waves (high frequency radio), and MICRO processors. Micro Design A/S promotes, sells, and supports the Q-Free products, using SAW, surface acoustic wave technology, as well as others. They support general AVI, Toll, Car Park, Railway, and Traffic control and management systems.

Micro Design A/S
P.O. Box 3974
N-7002, Tronelag, Norway

NDC AUTOMATION, INC.

NDC Automation, Inc. provides unique services to their customers. They perform the engineering and design services to meet the customer's needs, offering RF/ID systems to all types of manufacturing and material handling applications. Their wide range of applications include access control, parking lots, railroads, and pallet identification. Their focus is on automotive, food manufacturing, and pharmaceutical applications. They offer warehousing and distribution applications, and provide read-only and read/write RF/ID systems. The firm sells to end users, manufactures, OEM's, and systems integrators.

NDC Automation, Inc.
3101 Latrobe Dr.
Charlotte, NC 28211

NEDAP GIS

NEDAP GIS applications feature contact free identification with 4.3 billion unique codes available. There are read only and read/write capabilities available and they feature tool identification as a flexible aid to automation. Their application for manufacturing includes automatic tool and product identification. NEDAP offers a variety of tag types to be used for specialized application focusing on process automation uses, and vehicle identification for access control and positioning. The identification is used for tran-shipment control. NEDAP uses distributors for their end user support.

NEDAP GIS
P.O. Box 6, 7140 AA Groenio, Holland

SAVI TECHNOLOGY

Savi Technology emphasizes the system that automatically locates and identifies assets. They emphasize the asset management system using a radio technology to track your assets. The system uses small ID tags called the TyTag, to identify the asset. Fixed mounted or hand-held interrogator units are used to find the tags. Savi Asset Management Systems use a wide area technique to locate items with a cellular technology. Their applications manage worldwide movement of military equipment, container loading and unloading.

Savi Technology
260 Sheridan Avenue
Palo Alto, CA 94306

SENSORMATIC

Sensormatic makes automatic identification equipment used in retail sales. Their Sensor ID product comes in a variety of formats for electronic asset protection and tracking. They can track personnel, vehicles, and assets automatically. They produce an access control system. The Sensor ID can be mounted on walls, doorways, in floors and along conveyor belts. Sensormatic provides sales and support services in a world wide industrial marketplace. They have over 4,000 employees in 55 countries that install and maintain the systems. Sensormatic will use outside marketing but relies on internal installation support for their customers.

Sensormatic Electronics Corporation
500 Northwest 12th Avenue
Deerfield Beach, Florida 33442

SSI CUSTOM DATA CARDS

SSI Custom Data Cards manufactures multi-technology identification badges. Their emphasis is on security systems and other data collection systems. The technologies included in their product offerings are bar code, magnetic stripe, RF, and Weigan. SSI Custom Data Cards will custom fit the product to meet customer needs. They sell to distributors, end users, OEMs, VARs, and systems integrators.

SSI Custom Data Cards
1027 Waterwood Pkwy
Edmond, OK 73034

TELSOR CORP

Telsor Corporation makes passive RF/ID systems that are in the low frequency range. Telsor specializes in the short range identification. They produce both passive read-only and read/write systems without batteries. They feature accurate, industrial hardened, reliable products.

Telsor Corp.
6786 S. Revere Pkwy
Englewood, CO 80112

TEXAS INSTRUMENTS

Texas Instruments has made a significant investment in identification systems. They produce TIRIS. Texas Instruments manufactures RF/ID systems for use in automatic vehicle identification applications. They also have applications for security, work tracking and manufacturing automation. Their products and services feature read range, product durability, and reliability. The TIRIS products provide read-only and read/write versions. The products are offered direct to end users, manufacturers, and OEM. Texas Instruments also sells to systems integrators and VARS.

Texas Instruments, Inc.
34 Forest St., MS 20-27
Attleboro, MA 02703

VAPOR

Vapor designed and built, ROADCHECK (TM), Automatic Vehicle Identification equipment. They emphasize highway and railroad applications and offers upward compatibility from read-only to two-way communications. The Vapor system uses an approach where the antenna is in shadow slots in the pavement. Roadcheck's high data rate, 500 KBS, provides for five reads while a vehicle travels at 100 mph. The three components of the ROADCHECK system are the in-pavement antenna, the roadside reader, and the vehicle mounted transponder. Two way communications are available for applications such as toll roads and bridges, highways and turnpikes, airport, traffic monitoring, road pricing, dangerous goods movement, and so on. Vapor lists highway and railroad references with emphasis on advanced train control systems, ATCS.

Vapor Corporation
6420 West Howard Street
Chicago, Illinois 60648

X-CYTE INC.

X-Cyte produces automatic equipment identification systems using the surface acoustic wave technology. Their systems feature very low power levels and flexibility of system applications. Their early experience with toll applications and manufacturing systems using the RF/ID system lead to encryption and other security designs. X-Cyte holds patents in this technology. The application emphasis has been in the toll collection, access control, factory automation and vehicle ID applications. They use direct sales, as well as, dealers and distributors. They have an OEM program and consider VARs.

X-Cyte Inc.
2307 Bering Drive
San Jose, CA 95131

CHAPTER XIX

GLOSSARY OF TERMS

AAR	Association of American Railroads
AAR IR 173 AAR	TOFC/COFC interchange Rule 173 (1989).
AAR S-917-91	Specification for Application of Automatic Equipment Identification Transponders on Freight Cars.
AAR S-918-92	Standard for automatic equipment identification.
AASHTO	American Association of State Highway and Transportation Officials.
Active Tag	Transponder which has a power source of a battery or vehicle power, and continuously transmits a signal
AEI	Automatic Equipment Identification
AI	Artificial Intelligence, Rule based program to assist in the decision support process.
ANSI	American National Standards Institute
Antenna	A metallic apparatus for sending and receiving radio waves.
Antenna Beam Width	Generally interpreted as the 3dB beam width, or the space where power radiated from the antenna is within 3dB (or a factor of 0.5) of its maximum value.
APC	Automated Passenger Counting.

APTS	Advanced Public Transportation System- one system within IVHS.
ASK	Amplitude Shift Keying; A modulation technique in which the carrier wave is keyed on and off by the binary data signal.
ATA	American Trucking Association
ATC	Automated (electronic) Toll Collection.
ATIS	Advanced Traveler Information Systems. One system within IVHS.
ATMS	Advanced Traffic Management System. One system within IVHS.
ATSAC	Automatic Traffic Surveillance and Control.
Attenuate	To lower the power of the antenna signal.
AVC	Automatic Vehicle Classification, used in Toll and IVHS applications.
AVCS	Advanced Vehicle Control Systems, one system within IVHS.
AVI	Automatic Vehicle Identification; transponder based systems which monitor fixed locations.
Backscatter	Electromagnetic field reflected from objects in the field of view of a transmitting antenna.
BCS	Block Check Sequence.
CAD	Computer Aided Dispatch.

CALTRANS	The California Department of Transportation.
CCD	Charge Coupled Device, an optical electrical sensor.
Checksum	A code at the end of a frame that tells exactly how many bytes were transmitted.
CMS	Changeable Message Sign.
COFC	Container on Flat Car.
Container	Equipment used to transport objects, solids and liquids.
CRC	Cyclic Redundancy Check; is an error control method including data message and checking bits.
CVO	Commercial Vehicle Operation, an IVHS system.
CW	Continuous Wave.
Dead-Reckoning	A technique that calculates the current location of a vehicle by measuring the distance and direction that a vehicle has traveled since leaving a known starting point.
Differential Correction	A technique for overcoming GPS position determination errors.
DOT	Department of Transportation.
Dwell	The amount of time a vehicle takes to traverse a segment of a roadway. The vehicle may stop and wait and then move on. Used as a basis for fees at certain airports.
EDI	Electronic Data Interchange

EDP	Electronic Data Processing
EMF	Electro-Magnetic field.
EMI	Electro-Magnetic interference.
Emitter	Transponder
Encoder	A device used to program the identification code on the transponder.
ETC	Electronic Toll Collection
ETTM	Electronic Toll and Traffic Management
FCC	Federal Communications Commission
FHWA	Federal highway Administration.
Feature Extraction	used with photographs and images to identify the key features of the image as a way to determine important data about the image.
Frame	A time interval containing a complete signal.
Freight Container	Equipment used in international shipping, sits on a chassis or stack in a yard or on a ship, looks like a giant box, detailed in ISO 668.
FSK	Frequency Shift Keying; a digital modulation technique where the frequency is changed between two discrete values each generated by a separate oscillator embedded in the transponder.
FTA	Federal Transit Administration, part of U. S. DOT

GHz	Gigahertz; a measure of frequency, 1 billion cycles per second
GPS	Global Position Sensing. Sometimes called Global Positioning System. A government owned system of 24 earth orbiting satellites that transmit to ground based receivers. GPS provides extremely accurate latitude and longitude coordinates.
Handshakes	A term to describe the rate and amount of data transmission.
Hazmat	Hazardous material
HELP	Heavy Equipment License Plate program. A coalition of states.
HOV	High Occupancy Vehicle, e.g. bus, van
IBTTA	International Bridge, Tunnel and Turnpike Association
IC	Integrated Circuit.
IDEAS Program	Innovations Deserving Exploratory Analysis, a IVHS program to stimulate technology progress.
IEC	International Electrotechnical Commission
IEEE	Institute of Electrical and Electronics Engineers, Inc. A professional society.
IMO	Intermodal Marketing Organization
Inductive Loop	An antenna constructed of a wire loop embedded in the pavement that couples energy via a time varying magnetic field passing through the loop

Infrared	High end of the electromagnetic spectrum, just below visible light. Does not penetrate walls, etc. but instead will bounce off walls.
Interrogator	An antenna that first identifies or acknowledges a transponder as it enters a field of concern.
IR	Infrared.
ISO	International Standards Organization
ISO 6346	International Standards Organization, Freight Containers - Coding, Identification and Marking.
ISO 9001	International Standards Organization, Quality Systems
ISO DIS 10374	International Standards Organization, Standard for Automatic Identification of Containers
ISTEA	International Surface Transportation Act of 1991
IVHS	Intelligent Vehicle Highway System
IVHS America	Intelligent Vehicle Highway Society of America.
kbps	Kilobits Per Second; a measure of the speed of data flow, 1,000 bits per second.
kHz	Kilohertz; a measure of frequency, 1,000 cycles per second.
Manchester Coding	An encoding technique in which a binary one is represented by a transition from a logic one to a logic zero; a zero is a transition from a logic zero to a logic one.

Metering	A term used to control taxis and other vehicles at an airport. Set of queues.
MHz	Megahertz; a measure of frequency, 1 million cycles per second.
Micro	One millionth of a unit.
Microstrip Antenna	An antenna constructed of a path of metal deposited on a dielectric slab.
Microwave	High frequency electro-magnetic signal with a frequency between 300 MHZ and 300 GHz.
Mil Std 810D	Military standard, Environmental Test Methods and Engineering Guidelines.
Modulated Backscatter	Uses a principle of radio waves known as backscatter, or reflected energy, from a target toward the original illuminating source.
MIV	Mobile Inventory Vehicle. A vehicle with a radio frequency reader and a location system to determine the location of equipment in a yard.
Modulation	The process of varying the wave amplitude, frequency, or phase of a transmission signal.
Multiplex	Device used to control multiple antennas off one reader controller. Device gives control to each antenna.
Nano	One billionth of a unit.
Neural Networks	Computer systems that is rule based, uses routines patterned after human logic to determine answers. This is like a high powered artificial intelligence system. Corrections can be made by the system as the system learns from the input.

Passing speed	Speed at which a tag passes the sensing equipment.
Passive system	Utilizes a transponder which takes its power exclusively from the radiating antenna field.
Patch array	An array of microstrip antennas arranged in a two dimensional pattern.
Polarization	The particular state, either positive or negative, with reference to two poles or electrification. In most cases, a defined polarization or orientation of the transponder and reader antenna is needed to insure detection.
Power density	A measure of the radiated field intensity in watts per squared centimeter.
Range	Distance between the sensing equipment and the tag.
Read/Write	Transponder where two way communication exists.
RF	Radio Frequency; a method of wireless communication.
RF/ID	Radio Frequency Identification
SAW	Surface Acoustic Wave; uses piezoelectric materials, a pattern of electrodes on a surface, to generate an acoustic wave.
Semi-active System	Utilizes a transponder which has a self-contained battery. Usually the battery is activated when a signal reaches the transponder from a reader to assist in powering the reflected signal.

Shadowing — When vehicles follow each other closely through a lane of a toll plaza.

Smart Card — A contactless credit-card-size, card that uses a self- contained microprocessor and memory to store and process such items as toll fares, banking transactions, or identification. There are no metal contacts and this insures longer wear and resistance to problems caused by vibration.

Spectrum Analyzer — A tool used in transmission signal analysis. The peak signal frequency and amplitude variations are shown on a logarithmic scale. Simultaneously it measures voltages.

Tamper proof — Electronically Designed such that malicious modification of electronically stored information by subjection to electromagnetic signals from commonly available electronic devices is not possible.

Tamper proof — Physically Designed such that malicious disassembly, using commonly available tools, will be detected upon visual inspection.

TDMA — Time division multiple access; a protocol that uses time slices of different frequencies to communicate with multiple transponders in a field.

TOFC — Trailer on Flat Car.

Transceiver — A radio device that uses many of the same components for both transmission and reception.

Transponder — Small electronic device containing the alpha-numeric marking or identification code and related information; tag.

VES	Violation Enforcement System.
VRC	Vehicle Roadside Communications
Wiegan Power	Method of powering a tag using the incoming RF energy. This method can eliminate need for a battery.
WIM	Weigh- In -Motion
Yagi Antenna	A type of antenna, named for Professor H. Yagi; a series of horizontal rods decreasing in length from front to back. A commonly used TV antenna.
Zone	An area of an airport road and/or curb space designated for special use.

CHAPTER XX
FIGURES, CHARTS, TABLES, & PICTURES

CHAPTER XXI

INDEX